Youssef CHOUAIBI

General mechanics: Solid dynamics

Youssef CHOUAIBI

General mechanics: Solid dynamics

Course notes and corrected exercises for first-year mechanical engineering students

ScienciaScripts

Imprint

Any brand names and product names mentioned in this book are subject to trademark, brand or patent protection and are trademarks or registered trademarks of their respective holders. The use of brand names, product names, common names, trade names, product descriptions etc. even without a particular marking in this work is in no way to be construed to mean that such names may be regarded as unrestricted in respect of trademark and brand protection legislation and could thus be used by anyone.

Cover image: www.ingimage.com

This book is a translation from the original published under ISBN 978-620-6-71169-8.

Publisher:
Sciencia Scripts
is a trademark of
Dodo Books Indian Ocean Ltd. and OmniScriptum S.R.L publishing group

120 High Road, East Finchley, London, N2 9ED, United Kingdom
Str. Armeneasca 28/1, office 1, Chisinau MD-2012, Republic of Moldova, Europe
Printed at: see last page
ISBN: 978-620-8-27929-5

Table of contents

CHAPTER 1 : Geometry of masses

1. Centre of mass or centre of inertia

1.1. Definition

The centre of mass or centre of inertia of a system (Σ) is the point G in relation to which the mass is uniformly distributed. It is the barycentre of the points $M \in (\Sigma)$ affected by their mass dm. The position of the centre of mass G is determined using the following relationship:

$$\int \overrightarrow{GM}.\,dm = \vec{0} \tag{0.1}$$

We have

$$\overrightarrow{GM} = \overrightarrow{OM} - \overrightarrow{OG} \tag{0.2}$$

In this case :

$$\int \overrightarrow{OG}.\,dm = \int \overrightarrow{OM}.\,dm \tag{0.3}$$

The position of point G is independent of the mass parameters, hence :

$$\overrightarrow{OG}.\int dm = \int \overrightarrow{OM}.\,dm \tag{0.4}$$

If $\int dm = M$ where M is the mass of the system (Σ) then the position of the centre of mass G is given by :

$$M.\overrightarrow{OG} = \int \overrightarrow{OM}.\,dm \tag{0.5}$$

For a discrete system, the position of the centre of mass is calculated as follows:

$$m.\overrightarrow{OG} = \sum_{i=1}^{n} \overrightarrow{OG_i}.\,m_i \tag{0.6}$$

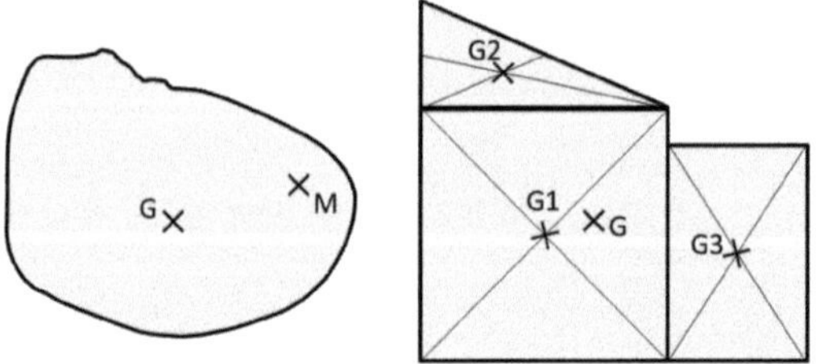

a) centre of mass for a continuous system b) centre of mass for a discrete system

Figure 0.1 Centre of mass for a system (Σ)

1.1.1. Material line

A material line is a material system with a very slender shape (wire, cable, etc.).

In this case the mass element "dm" can be modelled by :

$$dm = \lambda.\,dL \tag{0.7}$$

With λ : linear mass (kg/m). For a homogeneous solid λ =cts.

The mass of the system is $M = \lambda. L$ "where L is the total length. Using the definition (equation 1.5) we can write :

$$\lambda. L. \overrightarrow{OG} = \int \overrightarrow{OM}. \lambda. dL \rightarrow \quad L. \overrightarrow{OG} = \int \overrightarrow{OM}. dL \tag{0.8}$$

1.1.2. Material surface

If the material system (Σ) is a plane surface, then the mass element "dm" is in the following form :

$$dm = \sigma. ds \tag{0.9}$$

With σ: masse surfacique (kg/m^2)

In Cartesian coordinates : $ds = dx. dy$

In polar coordinates : $ds = r. dr. d\theta$

For a homogeneous solid "σ =cts" so the mass of the material system is " $M = \sigma. S$ "where 'S' is the total surface area. Referring to the definition (equation 1.5) we can write :

$$\sigma. S. \overrightarrow{OG} = \int \overrightarrow{OM}. \sigma. ds \rightarrow \quad S. \overrightarrow{OG} = \int \overrightarrow{OM}. ds \tag{0.10}$$

1.1.3. Material volume

If the hardware system (Σ) is a volume, then :

$$dm = \rho. dv \tag{0.11}$$

ρ: masse volumique $(kg/m)^3$

In Cartesian coordinates : $dv = dx. dy. dz$

In cylindrical coordinates : $dv = rdr. d\theta. dz$

In spherical coordinates : $dv = r^2. dr. \sin\theta. d\theta. d\varphi$

For a homogeneous solid "ρ=cts" so the mass " $m = \rho. V$ "where V is the total volume. Using the definition (equation 1.5) we can write :

$$\rho. V. \overrightarrow{OG} = \int \overrightarrow{OM}. \rho. dv \rightarrow \quad V. \overrightarrow{OG} = \int \overrightarrow{OM}. dV \tag{0.12}$$

1.2. Property of the centre of mass

1.2.1. Material symmetry

Material symmetry is geometric symmetry with symmetrical mass distributions. If a system has a point, axis or plane of symmetry, then its centre of gravity is located at that point, on that axis or in that plane of symmetry.

1.2.2. Calculation by operation

If a system (Σ) of mass M has several subsets, each with a mass m_i and a centre of mass Gi, then :

$$M. \overrightarrow{OG} = \sum_{i=1}^{n} \overrightarrow{OG_i}. m_i \tag{0.13}$$

$M. x_G = \sum_{i=1}^{n} x_{Gi}. m_i$;

$$M.y_G = \sum_{i=1}^{n} y_{Gi}.m_i \; ;$$

$$M.z_G = \sum_{i=1}^{n} z_{Gi}.m_i \; ;$$

1.3. GULDIN Theorem

GULDIN's Theorem offers an alternative approach for determining the position of the centre of inertia in certain specific cases without having to calculate integrals.

1.3.1. 1ère theorem

The lateral area produced by the rotation of a plane curve (L) about an axis Δ which lies in the same plane as (L) and does not intersect it, is equal to the product of the circumference generated by the centre of inertia G of curve (L) and the length L of curve (L).

Example: Rotating element 'dl' around axis Oy, without intercepting it, generates a surface 'ds' such that : $ds = 2.\pi.x_p.dl$. The total surface generated will be :

$$\mathbf{S = 2.\pi \int x_p . dl = 2.\pi.L.x_G} \tag{0.14}$$

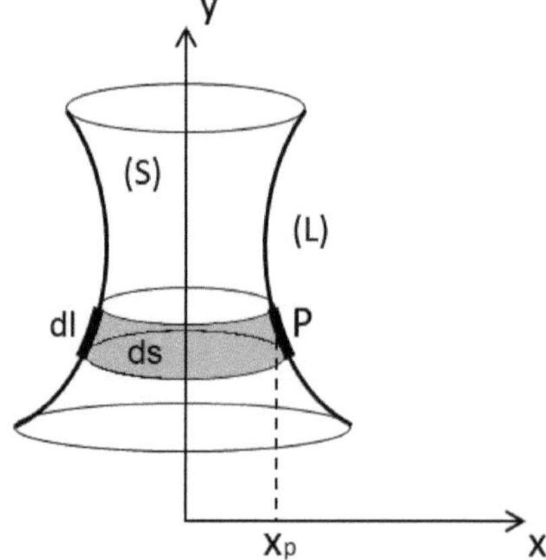

Figure 0.2 Centre of mass of a material line

1.3.2. 2 theorem

The volume created by the rotation of the plane surface S about an axis Δ which lies in the same plane as (S) and does not intersect it is equal to the product of the circumference generated by the point G which is the centre of inertia of the surface (S) and the area S of the surface (S).

Example: the rotation of the flat surface element 'ds' around the axis Oy, without intercepting it, generates a volume 'dv' such that : $dv = 2.\pi.x_p.ds$

The total volume generated will be :

$$V = 2.\pi.\int x_p.ds = 2.\pi.S.x_G \tag{0.15}$$

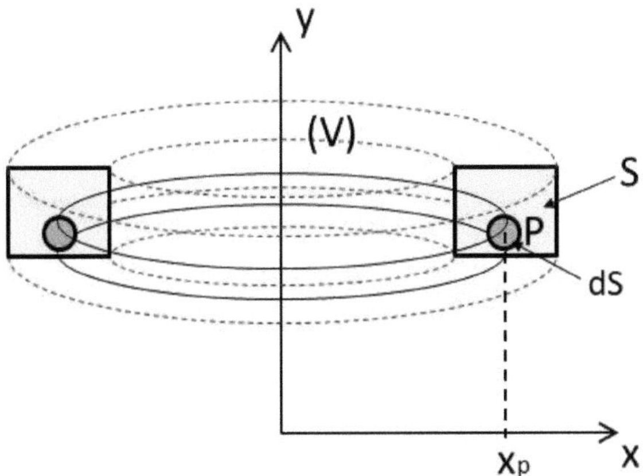

Figure 0.3 Centre of mass of a surface

1.4. Applications

<u>**Example 1**</u>:

Consider a quarter circle of radius R with linear massλ and length L.

1- Use the definition of a material line to find the centre of inertia G with coordinates (y_G , z_G)

2- Use the 1er GULDIN theorem to determine the coordinates (y_G , z_G) of the centre of inertia G.

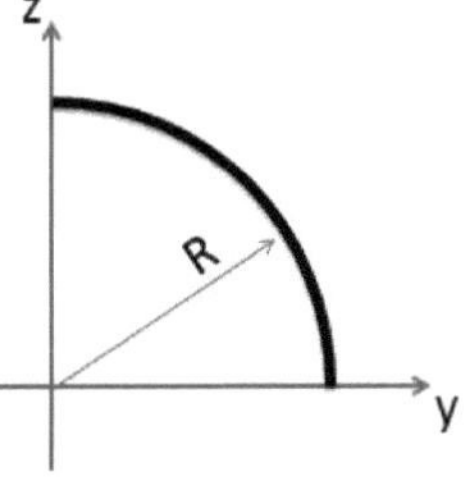

Correction:

1- The definition of the centre of mass of the material line gives :

$$\lambda.L.\overrightarrow{OG} = \int \overrightarrow{OM}.\lambda.dL \rightarrow \quad L.\overrightarrow{OG} = \int \overrightarrow{OM}.dL$$

$$L = \int_0^L dL = \int_0^{\frac{\pi}{2}} R.d\theta = R.\frac{\pi}{2}$$

$$L.y_G = \int_0^{\frac{\pi}{2}} R.\cos(\theta).R.d\theta \rightarrow \quad R.\frac{\pi}{2}.y_G = R^2.[\sin(\theta)]_0^{\frac{\pi}{2}} = R^2$$

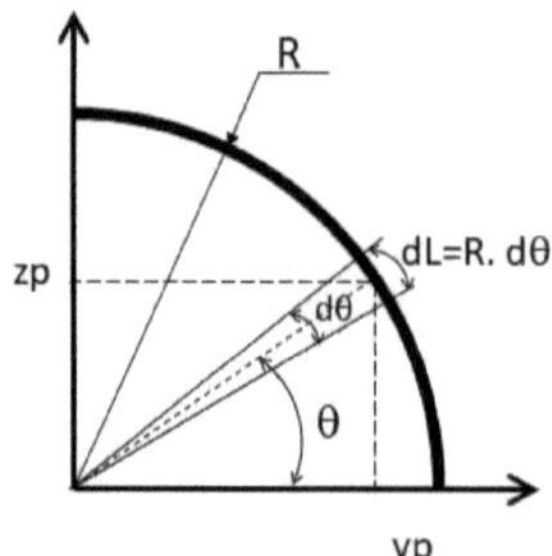

$$y_G = \frac{2.R}{\pi} \; ; Z_G = \frac{2.R}{\pi}$$

2- Use the 1er GULDIN theorem

$$S = 2.\pi \int y_p \,.\, dl = 2.\pi.\,L.\,y_G$$

$$S = 2.\pi \int y_p \,.\, dl = 2.\pi \int_0^{\frac{\pi}{2}} R.\cos(\theta)\,.\,R.\,d\theta = 2.\pi.\,R^2$$

$$S = 2.\pi.\,R^2 = 2.\pi.\,L.\,y_G = \pi^2.\,R.\,y_G \rightarrow \; y_G = \frac{2.R}{\pi}$$

Example 2:

Consider a quarter disc with radius R, mass per unit area σ and surface area S.

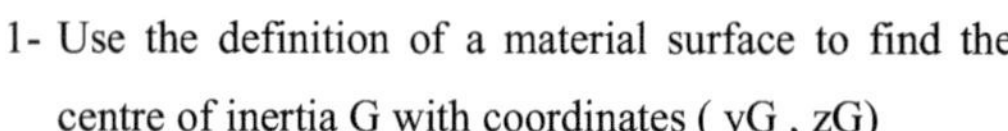

1- Use the definition of a material surface to find the centre of inertia G with coordinates (y_G , z_G)

2- Use the $2^{\text{ème}}$ GULDIN theorem to determine the coordinates (y_G , z_G) of the centre of inertia G.

Correction

1- Using the definition of the centre of mass of a material surface :

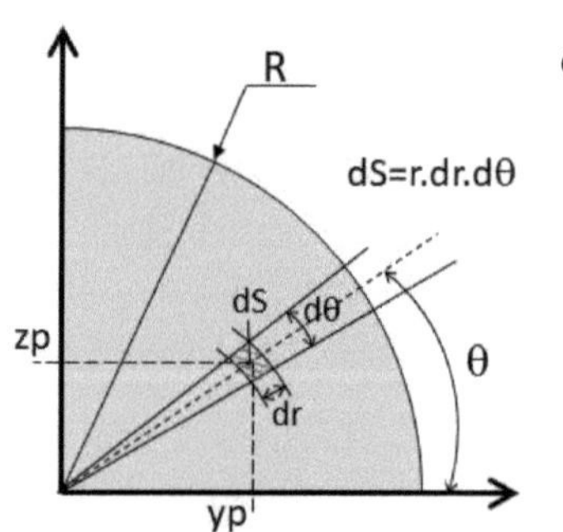

$$S.\overrightarrow{OG} = \int \overrightarrow{OM}.\,ds$$

$$S = \int ds = \int_0^{\frac{\pi}{2}} \int_0^{R} r.\,dr.\,d\theta = \pi \frac{R^2}{4}$$

$$S.y_G = \int_0^{\frac{\pi}{2}} \int_0^{R} r^2.\cos(\theta)\,.\,dr.\,d\theta \rightarrow \; \pi \frac{R^2}{4}.\,y_G = \frac{R^3}{3}$$

$$y_G = \frac{4.R}{3.\pi} \; ; Z_G = \frac{4.R}{3.\pi}$$

2- Use the $2^{\text{ème}}$ GULDIN theorem

$$V = 2.\pi.\int y_p.\,ds = 2.\pi.\,S.\,y_G$$

$$V = 2.\pi \int y_p.\,dS = 2.\pi \int_0^{\frac{\pi}{2}} \int_0^{R} r^2.\cos(\theta)\,.\,dr.\,d\theta = 2.\pi.\frac{R^3}{3}$$

$$2.\pi.\frac{R^3}{3} = 2.\pi.\,S.\,y_G = 2.\pi.\frac{\pi.R^2}{4}.\,y_G \rightarrow \; y_G = \frac{4.R}{3.\pi} \; ; Z_G = \frac{4.R}{3.\pi}$$

2. Moment of inertia

2.1. Definition

The moment of inertia is a physical quantity that characterises the distribution of mass within a solid (Σ). It represents the resistance of a solid (Σ) to being rotated. It is difficult to rotate the system (Σ) if the moment of inertia becomes greater and greater.

The moment of inertia of the system (Σ) with respect to a reference element **"r"** is the scalar defined by :

$$I_O(\Sigma) = \int \overrightarrow{OM}^2 dm \tag{0.16}$$

$$I_\Delta(\Sigma) = \int \overrightarrow{\Delta M}^2 dm \tag{0.17}$$

$$I_\pi(\Sigma) = \int \overrightarrow{\pi M}^2 dm \tag{0.18}$$

The reference element **"r"** can be either a point (O), a straight line (Δ) or a plane (π).

$OM, \Delta M$ and πM are the minimum distances from point M to point O, to the straight line Δ and to the plane π.

2.2. Expression of moments of inertia relative to the reference frame $R(O, \vec{x}, \vec{y}, \vec{z})$

- In relation to the planes of R :

$$I_{Oxy}(\Sigma) = \int_{M \in \Sigma}^{\cdot} z^2 dm \tag{0.19}$$

$$I_{Oyz}(\Sigma) = \int_{M \in \Sigma}^{\cdot} x^2 dm \tag{0.20}$$

$$I_{Oxz}(\Sigma) = \int_{M \in \Sigma}^{\cdot} y^2 dm \tag{0.21}$$

- In relation to the axes of R :

$$I_{Ox}(\Sigma) = A = \int_{M \in \Sigma}^{\cdot} (y^2 + z^2) dm \tag{0.22}$$

$$I_{Oy}(\Sigma) = B = \int_{M \in \Sigma}^{\cdot} (x^2 + z^2) dm \tag{0.23}$$

$$I_{Oz}(\Sigma) = C = \int_{M \in \Sigma}^{\cdot} (x^2 + y^2) dm \tag{0.24}$$

- Relative to the centre of R :

$$I_O(\Sigma) = \int_{M \in \Sigma}^{\cdot} (x^2 + y^2 + z^2) dm \tag{0.25}$$

- Product of inertia :

$$J_{xOy}(\Sigma) = \int_{M \in \Sigma}^{\cdot} x.y \, dm \tag{0.26}$$

$$J_{xOz}(\Sigma) = \int_{M \in \Sigma}^{\cdot} x.z \, dm \tag{0.27}$$

$$J_{yOz}(\Sigma) = \int_{M \in \Sigma}^{\cdot} y.z \, dm \tag{0.28}$$

2.3. Relationship between the different moments of inertia

1 relationship :

$$I(O,\vec{x}) = I(O,\vec{x},\vec{y}) + I(O,\vec{x},\vec{z}) \tag{0.29}$$

$$I(O,\vec{y}) = I(O,\vec{x},\vec{y}) + I(O,\vec{y},\vec{z}) \tag{0.30}$$

$$I(O,\vec{z}) = I(O,\vec{x},\vec{z}) + I(O,\vec{y},\vec{z}) \tag{0.31}$$

2 relationship :

$$I_O = I(O,\vec{x},\vec{y}) + I(O,\vec{x},\vec{z}) + I(O,\vec{y},\vec{z}) \tag{0.32}$$

$$I_O = \frac{1}{2}[I(O,\vec{x}) + I(O,\vec{y}) + I(O,\vec{z})] \tag{0.33}$$

3 relationship :

$$I(O,\vec{x}) + I(O,\vec{y}) = I(O,\vec{z}) + 2.I(O,\vec{x},\vec{y}) \tag{0.34}$$

$$I(O,\vec{y}) + I(O,\vec{z}) = I(O,\vec{x}) + 2.I(O,\vec{y},\vec{z}) \tag{0.35}$$

$$I(O,\vec{x}) + I(O,\vec{z}) = I(O,\vec{y}) + 2.I(O,\vec{x},\vec{z}) \tag{0.36}$$

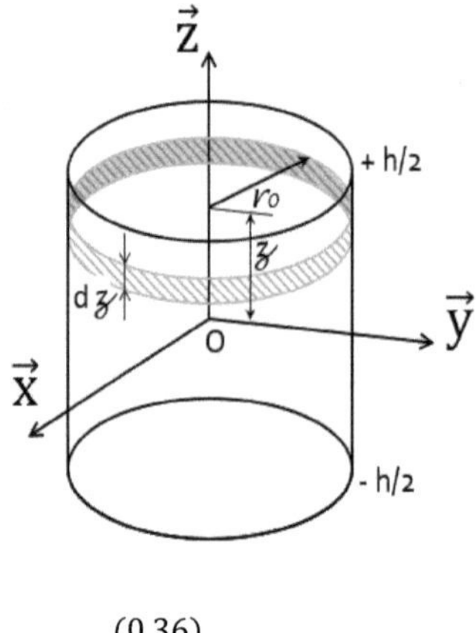

2.4. Applications

Consider a homogeneous solid cylinder with radius r_0 and height h. The origin of the reference frame is centred at G, the cylinder's centre of mass, and the axis Oz coincides with the cylinder's axis.

1- Calculate the moment of the cylinder relative to the Oxy plane.

2- Then calculate the cylinder's moment of inertia about the Oz axis.

<u>***Solution:***</u>

1-

The density ρ is constant, so we can write :

$dm = \rho dV = \rho(\pi r_0{}^2)dz$ ($m = \rho\pi r_0{}^2 h$ where m is the total mass of the cylinder)

$$I_{Oxy}(\Sigma) = \int_{-h/2}^{+h/2} z^2 dm = \int_{-h/2}^{+h/2} z^2 \rho(\pi r_0{}^2)dz$$

$$I_{Oxy}(\Sigma) = \rho.\pi.r_0{}^2 \int_{-h/2}^{+h/2} z^2 dz = \rho.\pi.r_0{}^2.\frac{1}{3}\left(\frac{h^3}{8}+\frac{h^3}{8}\right)$$

$$I_{Oxy}(\Sigma) = \rho.\pi.r_0{}^2.\frac{h^3}{12} = m.\frac{h^2}{12}$$

2-

Let's apply the definition :

$$I_{Oz}(\Sigma) = \int_{M\in\Sigma} (x^2+y^2)dm$$

$$= \int_{M\in\Sigma} r^2 dm$$

$dm = \rho dV = \rho r dr d\theta dz = 2\pi h \rho r dr$

Hence

$$I_{Oz}(\Sigma) = \int_0^{r_0} r^2.2\pi h\rho r.dr$$

$$= 2\pi h\rho \int_0^{r_0} r^3 dr$$

$$I_{Oz}(\Sigma) = 2\pi h\rho \frac{r_0{}^4}{4} = m\frac{r_0{}^2}{2}$$

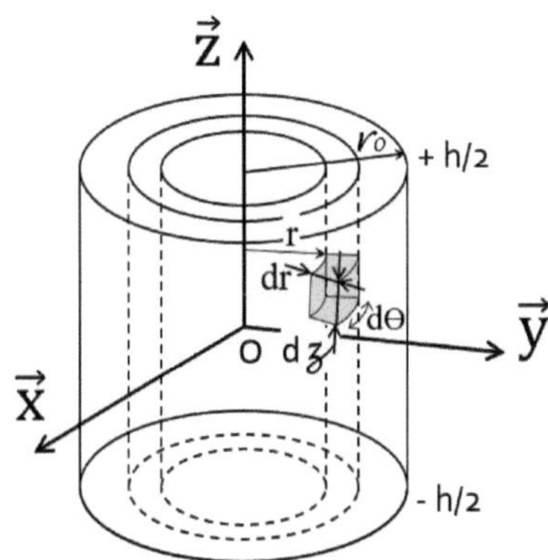

3. Operators or inertia matrix

3.1. Definition

The linear application denoted by $J_O(S)$ where : $J_O(S): \mathbb{R}^3 \to \mathbb{R}^3$

$$\vec{u} \longmapsto J_O(S)(\vec{u}) = \int \overrightarrow{OM} \wedge (\vec{u} \wedge \overrightarrow{OM}).\,dm \qquad (0.37)$$

3.2. Inertia matrix

Associated with $J_O(S)$ a matrix called the inertia matrix of (S) relative to the point O is noted $[I_O(s)]$ such that its columns are the images of $(\vec{x}, \vec{y}, \vec{z})$ by $J_O(S)$.

$$[I_O(s)] = [J_O(S)(\vec{x});\ J_O(S)(\vec{y});\ J_O(S)(\vec{z})] \qquad (0.38)$$

$$[I_0(S)] = \begin{bmatrix} A & -F & -E \\ -F & B & -D \\ -E & -D & C \end{bmatrix} \qquad (0.39)$$

The elements in the first column of the inertia matrix represent the components of the vector

$$J_0(S)(\vec{x}) = \int \overrightarrow{OM} \wedge (\vec{x} \wedge \overrightarrow{OM}).\,dm \qquad (0.40)$$

Let's ask

$$\overrightarrow{OM} = x.\vec{x} + y.\vec{y} + z.\vec{z} \qquad (0.41)$$

So

$$\vec{x} \wedge \overrightarrow{OM} = \begin{pmatrix} 1 \\ 0 \\ 0 \end{pmatrix} \wedge \begin{pmatrix} x \\ y \\ z \end{pmatrix} = \begin{pmatrix} 0 \\ -z \\ y \end{pmatrix} \qquad (0.42)$$

$$\overrightarrow{OM} \wedge (\vec{x} \wedge \overrightarrow{OM}) = \begin{pmatrix} x \\ y \\ z \end{pmatrix} \wedge \begin{pmatrix} 0 \\ -z \\ y \end{pmatrix} = \begin{pmatrix} y^2 + z^2 \\ -x.y \\ -x.z \end{pmatrix} \qquad (0.43)$$

$$A = \int (y^2 + z^2).\,dm \qquad (0.44)$$

$$F = \int (x.y).\,dm \qquad (0.45)$$

$$E = \int (x.z).\,dm \qquad (0.46)$$

After an identical calculation, we can determine the other two columns. We find :

$$B = \int (x^2 + z^2).\,dm \qquad (0.47)$$

$$C = \int (x^2 + y^2).\,dm \qquad (0.48)$$

$$D = \int (y.z).\,dm \qquad (0.49)$$

3.3. Influence of symmetry on the inertia matrix

To obtain the simplest matrix, we need to choose the base and the point at which we will express the inertia matrix, taking account of material symmetries.

+Material symmetry

+Geometric symmetry

+Symmetry of material distribution

3.3.1. Material symmetry in relation to a plane

(S) has a plane of symmetry $(o, \vec{y}, \vec{z})$ $P(x,y,z)$ is associated with a point $P'(-x,y,z)$ symmetrical about $(o, \vec{y}, \vec{z})$. Let (S+) be the part of (S) such that x>=0 and (S-) be the part of (S) such that x<=0 then

$$S = S^+ \cup S^- \tag{0.50}$$

$$E = \iiint_{(S)} xz \, . \, dm = \iiint_{(S+)} xz \, . \, dm + \iiint_{(S-)} xz \, . \, dm \tag{0.51}$$

Consider the symmetry

$$\iiint_{(S+)} xz \, . \, dm = - \iiint_{(S-)} xz \, . \, dm \tag{0.52}$$

$$E = 0 \tag{0.53}$$

$$F = \iiint xy \, . \, dm = 0 \tag{0.54}$$

So

$$[I_0(S)] = \begin{bmatrix} A & 0 & 0 \\ 0 & B & -D \\ 0 & -D & C \end{bmatrix} \tag{0.55}$$

A: principal moment of inertia the axis $(o, \vec{x})$ is a principal axis of inertia

Note: if (S) is a plate in $(o, \vec{y}, \vec{z})$, then $A = B + C$

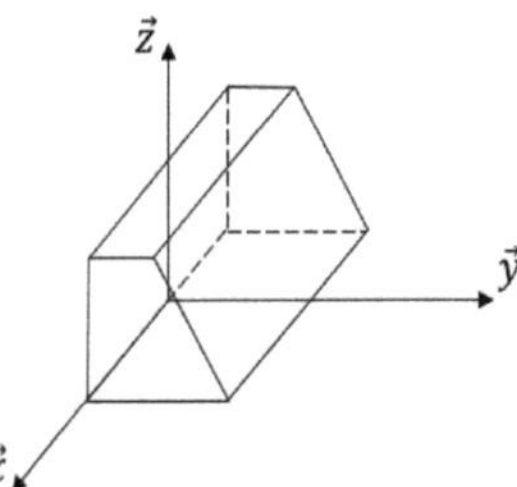

Figure 0.4 Material symmetry with respect to a plane

3.3.2. Material symmetry in relation to an axis of revolution

(S) has two planes of symmetry $(o, \vec{y}, \vec{z})$ and $(o, \vec{x}, \vec{z})$. All products of inertia are zero.

The axes $(o, \vec{x})$ and $(o, \vec{y})$ are equivalent, so $A = B$

$$A = \iiint (y^2 + z^2) \, . \, dm = \iiint (x^2 + z^2) \, . \, dm = B \tag{0.56}$$

$$[I_0(S)] = \begin{bmatrix} A & 0 & 0 \\ 0 & A & 0 \\ 0 & 0 & C \end{bmatrix} \qquad (0.57)$$

Figure 0.5 Material symmetry with respect to an axis of revolution

3.3.3. Material symmetry with respect to a point :

All the planes and axes of the reference frame are planes and axes of symmetry.

$$[I_0(S)] = \begin{bmatrix} A & 0 & 0 \\ 0 & A & 0 \\ 0 & 0 & A \end{bmatrix} \qquad (0.58)$$

Figure 0.6 Material symmetry with respect to a point

Note:
- When a solid has a plane of material symmetry, the principal axis of inertia is an axis perpendicular to this plane.
- The inertia matrix is principal if a solid has two orthogonal planes of material symmetry.
- When a solid has two perpendicular axes of symmetry, then the inertia matrix is principal
- Any axis of symmetry is a principal axis of inertia

3.3.4. Applications

Example 1:

Consider a quarter disc of radius R, mass per unit area σ

and surface area S.

1- Find the inertia matrix $[I_0(S)]$

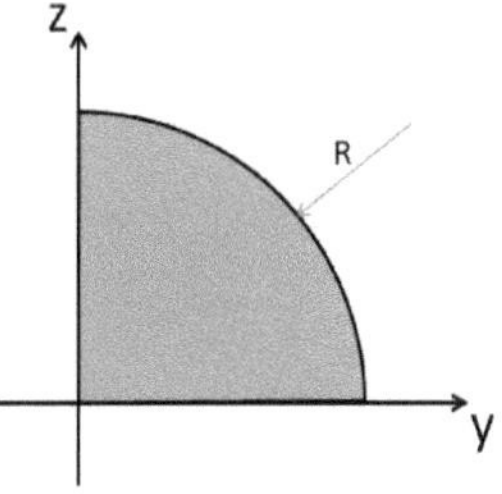

__Correction__

$(O, \vec{Z})$ is a principal axis of inertia is principal then the products of inertia D=E=0

The axes **Ox** and **Oy** are equivalent so : A=B

$$[I_0(S)] = \begin{bmatrix} A & -F & 0 \\ -F & B & 0 \\ 0 & 0 & C \end{bmatrix}$$

$$A = \int (x^2 + z^2).\,dm \quad (1)$$
$$B = \int (y^2 + z^2).\,dm \quad (2)$$
$$C = \int (x^2 + y^2).\,dm \quad (3)$$

The disc is of negligible thickness (z=0) so :

$$A = \int (x^2).\,dm \quad (4)$$
$$B = \int (y^2).\,dm \quad (5)$$
$$C = \int (x^2 + y^2).\,dm \quad (6)$$

The moment of inertia along the oz axis is the sum of the two moments of inertia :

$$A = B = C/2$$

$$C = \int (x^2 + y^2).\,dm = \int r^2.\,dm = \sigma \int r^2.\,ds = \sigma \int_0^{\frac{\pi}{2}} \int_0^R r^2.\,r.\,dr.\,d\theta$$

$$C = \sigma.\frac{\pi}{2}.\frac{R^4}{4} = m.\frac{R^2}{2}$$

$$A = B = m.\frac{R^2}{4}$$

$$F = \int xy.\,dm = \sigma \int xy.\,ds$$

$$F = \sigma \int_0^{\frac{\pi}{2}} \int_0^R r.\cos(\theta).r.\sin(\theta).r.\,dr.\,d\theta$$

$$F = \sigma \int_0^{\frac{\pi}{2}} \cos(\theta).\sin(\theta).\,d\theta \int_0^R .r^3.\,dr$$

$$F = \sigma \frac{R^4}{4} \int_0^{\frac{\pi}{2}} \cos(\theta).\sin(\theta).\,d\theta$$

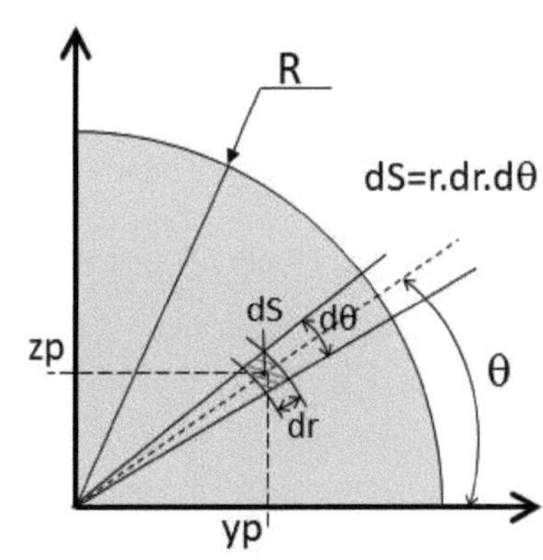

$$F = \sigma \frac{R^4}{4} \int_0^{\frac{\pi}{2}} . \sin(\theta) . d(\sin(\theta))$$

$$F = \sigma \frac{R^4}{4} \left[\frac{1}{2} \sin(\theta)^2 \right]_0^{\frac{\pi}{2}} = \sigma \frac{R^4}{4} = m \frac{R^2}{2.\pi}$$

$$[I_0(S)] = \begin{bmatrix} \dfrac{m.R^2}{4} & -\dfrac{m.R^2}{2.\pi} & 0 \\[2mm] -\dfrac{m.R^2}{2.\pi} & \dfrac{m.R^2}{4} & 0 \\[2mm] 0 & 0 & \dfrac{m.R^2}{2} \end{bmatrix}$$

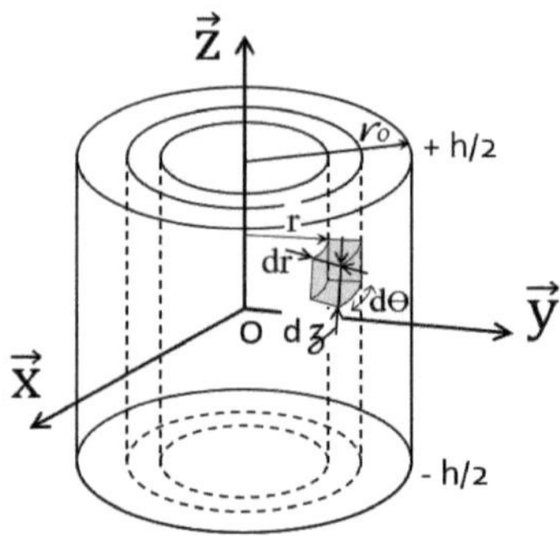

Example 2

2- Let (S) be a homogeneous cylinder of radius R and height h: Find the inertia matrix $[I_0(S)]$

$(O,\vec{Z})$ is an axis of revolution (axis of material symmetry) then the inertia matrix is principal

$$[I_0(S)] = \begin{bmatrix} A & 0 & 0 \\ 0 & A & 0 \\ 0 & 0 & C \end{bmatrix}$$

$$A = \int (x^2 + z^2).dm \quad (1)$$
$$A = \int (y^2 + z^2).dm \quad (2)$$
$$C = \int (x^2 + y^2).dm \quad (3)$$

Let's add member to member the relations (1) and (2)

$$2.A = \int (x^2 + y^2).dm + 2.\int z^2.dm$$

This reveals C. from which

$$A = \frac{C}{2} + \int z^2.dm \quad \text{with } dm = \rho.r.dr.d\theta.dz$$

$$dm = \rho.\pi.R^2.dz = \frac{m}{h}dz \quad \text{hence} \quad \int z^2.dm = \frac{m}{h}.\int z^2.dz = \frac{m.h^2}{12}$$

$$C = \int (x^2 + y^2).dm = \int r^2.dm = \frac{2.m}{R^2}\int r^3.dr = \frac{m.R^2}{2}$$

$$A = \frac{C}{2} + \int z^2.dm = \frac{m.R^2}{4} + \frac{m.h^2}{12}$$

3.4. HUYGENS theorem

Huygens' theorem is a mathematical formula that establishes a relationship between the inertia matrices at a point G, the centre of mass of a body, and at any point M.

Let G be the centre of inertia of (S), $R_G(G, \vec{x}_1, \vec{y}_1, \vec{z}_1)$ is an ROD such that $\vec{x}_1 /\!/ \vec{x}$, $\vec{y}_1 /\!/ \vec{y}$, $\vec{z}_1 /\!/ \vec{z}$

Let M be a point belonging to (S) and the coordinates of M in the reference frame RG are

$M(x \quad y \quad z)_{(\overline{x},\overline{y},\overline{z})}$ Let $G(x_G \quad y_G \quad z_G)_{(\overline{x},\overline{y},\overline{z})}$ and $M(x_1 \quad y_1 \quad z_1)_{(\vec{x}_1,\vec{y}_1,\vec{z}_1)}$ then

$M(x_1 + x_G \quad y_1 + y_G \quad z_1 + z_G)_{(\overline{x},\overline{y},\overline{z})}$

$$A = \iiint (y^2 + z^2).\,dm = \iiint (y_1 + y_G)^2 + (z_1 + z_G)^2.\,dm \tag{0.59}$$

$$A = \iiint y_1{}^2 + z_1{}^2.\,dm + (y_G{}^2 + z_G{}^2)\iiint dm + 2.y_G \iiint z_1\, dm + 2.z_G \iiint y_1\, dm \tag{0.60}$$

$$\iiint z_1\, dm = \iiint y_1\, dm = 0 \tag{0.61}$$

$$A = I(G, \vec{x}_1)(S) + (y_G{}^2 + z_G{}^2).\,m \tag{0.62}$$

So

$$A = I(O, \vec{x}) = I(G, \vec{x}_1)(S) + m.\,(y_G{}^2 + z_G{}^2) \tag{0.63}$$

$$B = I(O, \vec{y}) = I(G, \vec{y}_1)(S) + m.\,(x_G{}^2 + z_G{}^2) \tag{0.64}$$

$$C = I(O, \vec{z}) = I(G, \vec{z}_1)(S) + m.\,(x_G{}^2 + y_G{}^2) \tag{0.65}$$

$$D = I_{yz} = \int_{(S)} yz\, dm = \int_{(S)} y_1 z_1\, dm + y_G z_G.\int_{(S)} dm + y_G.\int_{(S)} z_1\, dm + z_G.\int_{(S)} y_1\, dm \tag{0.66}$$

$$I_{yz} = I_G yz + m y_G z_G \tag{0.67}$$

$$I_{xy} = I_G xy + m x_G y_G \tag{0.68}$$

$$I_{xz} = I_G xz + m x_G z_G \tag{0.69}$$

$$\begin{bmatrix} A & -F & -E \\ -F & B & -D \\ -E & -D & C \end{bmatrix} = \begin{bmatrix} A_G & -F_G & -E_G \\ -F_G & B_G & -D_G \\ -E_G & -D_G & C_G \end{bmatrix} + m \begin{bmatrix} y_G{}^2 + z_G{}^2 & -x_G y_G & -x_G z_G \\ -x_G y_G & x_G{}^2 + z_G{}^2 & -y_G z_G \\ -x_G z_G & -y_G z_G & x_G{}^2 + y_G{}^2 \end{bmatrix} \tag{0.70}$$

$$[I_0(S)] = [I_G(S)] + [I(G.\,O.\,m)] \tag{0.71}$$

$[I(G.\,O.\,m)]$ Matrix of inertia of point G affected by the mass relative to point O.

Caution: This relationship only allows you to pass immediately through the centre of mass G of the solid.

3.5. Case of a complex solid composed of elementary solids

In some cases, it may be useful to divide a solid into simpler elements, for which the inertia matrices are easy to determine or already known.

$$[I_0(S)] = [I_0(S1)] + [I_0(S2)] - [I_0(S3)] \tag{0.72}$$

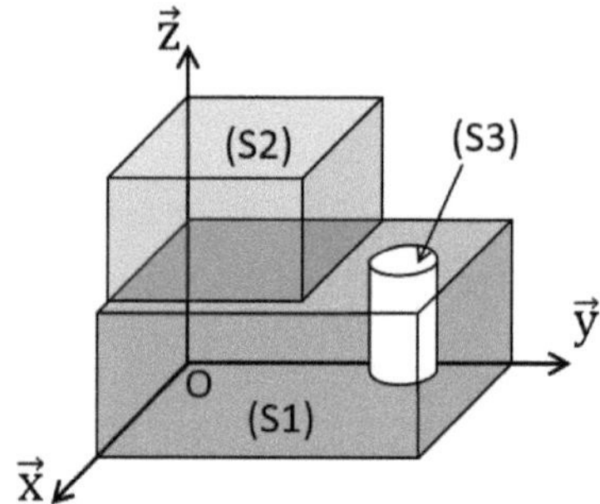

Figure 0.7 Inertia matrix of a complex solid

4. Application:

4.1. Exercise 1: Inertia matrix of a crank handle

The crank (S) shown in the figure below consists of three cylindrical bodies.

Find the inertia matrix $[I_0(S)]$

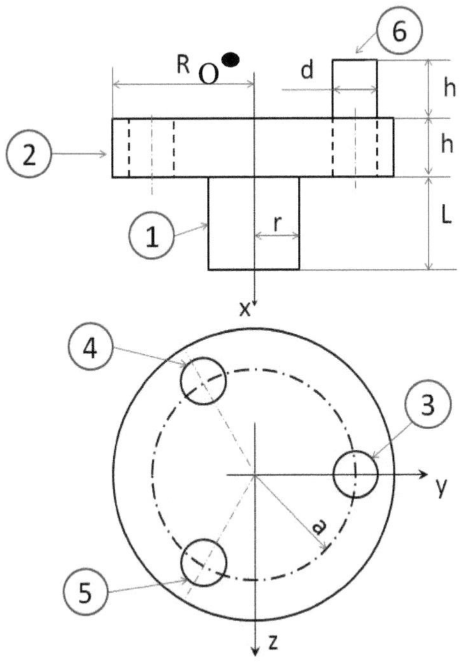

Figure 0.8 Inertia matrix of a crank handle

$$[I_0(S)] = [I_0(S1)] + [I_0(S2)] + [I_0(S6)] - [I_0(S3)] - [I_0(S4)] - [I_0(S5)]$$

$$[I_G(Si)] = \begin{bmatrix} \dfrac{mi.\,R^2}{2} & 0 & 0 \\[2mm] 0 & \dfrac{mi.\,R^2}{4} + \dfrac{mi.\,h^2}{12} & 0 \\[2mm] 0 & 0 & \dfrac{mi.\,R^2}{4} + \dfrac{mi.\,h^2}{12} \end{bmatrix}$$

$$[I_0(Si)] = [I_{Gi}(Si)] + [I(Gi.\,O.\,mi)]$$

4.2. Exercise 2: aluminium plate with holes.

Consider a square plate with a cylindrical hole of radius R. The centre of the hole is offset by a distance d along the y axis. The centre of the hole is offset by a distance d along the y axis (see Figure 1.8).

We give :

$\rho = 2700$ kg/m3; R = 0.6 m; d = 1 m; L = 4 m and e = 0.1 m

1- Calculate the coordinates of the centre of mass of the plate in the base $(\vec{x}, \vec{y}, \vec{z})$.

2- Determine the inertia matrix of the plate with the hole at point O : $[I_O(S)]$

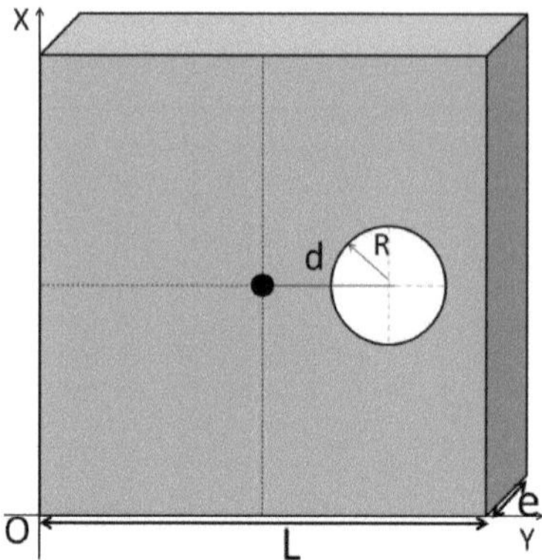

Figure 0.9 Square plate with holes

Correction:

1 / The position of the centre of mass G is given by : $M.\overrightarrow{OG} = \int \overrightarrow{OM}.\,dm$

If the plate is a flat surface, then the mass element "dm" is in the following form: dm=ρ .dv

With ρ: masse volumique (kg/m^3) For a homogeneous solid "ρ =cts" so the mass of the material system is " $M = V.S$ "where 'V' is the total volume. Using the definition of the centre of mass, we can write :

$$V.\rho.\overrightarrow{OG} = \int \overrightarrow{OM}.\rho.\,dV \rightarrow \qquad V.\overrightarrow{OG} = \int \overrightarrow{OM}.\,dV$$

$$V = \int dV = \int_0^L \int_0^L \int_0^e dx.\,dy\,dz = L^2 e$$

So $V.\overrightarrow{OG} = \int \overrightarrow{OM}.\,dV \rightarrow L^2e.\overrightarrow{OG} = \int_0^L \int_0^L \int_0^e \begin{pmatrix} X \\ y \\ z \end{pmatrix} dx.\,dy\,dz = \begin{pmatrix} \dfrac{L^3}{2}.e \\ \dfrac{L^3}{2}.e \\ L^2 \dfrac{e^2}{2} \end{pmatrix}$

The centre of mass is $\overrightarrow{OG} = \begin{pmatrix} \dfrac{L}{2} \\ \dfrac{L}{2} \\ \dfrac{e}{2} \end{pmatrix}$

2/ (S) has two planes of symmetry $(o, \vec{y}, \vec{z})$ and $(o, \vec{x}, \vec{z})$, so D=E=F=0.

The axes $(o, \vec{x})$ and $(o, \vec{y})$ are equivalent, so $A = B$

$$[I_G(S)] = \begin{bmatrix} A & -F & -E \\ -F & B & -D \\ -E & -D & C \end{bmatrix} = \begin{bmatrix} A & 0 & 0 \\ 0 & A & 0 \\ 0 & 0 & C \end{bmatrix}$$

$$A = B = \int (y^2 + z^2).dm = \rho \int_{-\frac{L}{2}}^{\frac{L}{2}} \int_{-\frac{L}{2}}^{\frac{L}{2}} \int_{-\frac{e}{2}}^{\frac{e}{2}} (y^2 + z^2) dx.\,dy\,dz = \rho \left(\int_{-\frac{L}{2}}^{\frac{L}{2}} \int_{-\frac{L}{2}}^{\frac{L}{2}} \int_{-\frac{e}{2}}^{\frac{e}{2}} y^2 dx.\,dy\,dz + \right.$$

$$\left. \int_{-\frac{L}{2}}^{\frac{L}{2}} \int_{-\frac{L}{2}}^{\frac{L}{2}} \int_{-\frac{e}{2}}^{\frac{e}{2}} z^2 dx.\,dy\,dz \right) = \rho \left(\frac{L^3}{12}.L.\,e + L^2.\frac{e^3}{12} \right) = \frac{m}{12} (L^2 + e^2)$$

$$C = \int (x^2 + y^2).dm = \rho \int_{-\frac{L}{2}}^{\frac{L}{2}} \int_{-\frac{L}{2}}^{\frac{L}{2}} \int_{-\frac{e}{2}}^{\frac{e}{2}} (x^2 + y^2) dx.\,dy\,dz$$

$$C = \rho \left(\int_{-\frac{L}{2}}^{\frac{L}{2}} \int_{-\frac{L}{2}}^{\frac{L}{2}} \int_{-\frac{e}{2}}^{\frac{e}{2}} x^2 dx.\,dy\,dz + \int_{-\frac{L}{2}}^{\frac{L}{2}} \int_{-\frac{L}{2}}^{\frac{L}{2}} \int_{-\frac{e}{2}}^{\frac{e}{2}} y^2 dx.\,dy\,dz \right) = \rho \left(\frac{L^3}{12}.L.\,e + \frac{L^3}{12}.L.\,e \right) = m \frac{L^2}{6}$$

$$[I_G(S)] = \begin{bmatrix} \frac{m}{12}(L^2 + e^2) & 0 & 0 \\ 0 & \frac{m}{12}(L^2 + e^2) & 0 \\ 0 & 0 & m\frac{L^2}{6} \end{bmatrix}$$

$$[I_O(S)] = \begin{bmatrix} A_G & 0 & 0 \\ 0 & B_G & 0 \\ 0 & 0 & C_G \end{bmatrix} + m \begin{bmatrix} y_G{}^2 + z_G{}^2 & -x_G y_G & -x_G z_G \\ -x_G y_G & x_G{}^2 + z_G{}^2 & -y_G z_G \\ -x_G z_G & -y_G z_G & x_G{}^2 + y_G{}^2 \end{bmatrix}$$

$$[I_O(S)] = \begin{bmatrix} \frac{m}{12}(L^2 + e^2) & 0 & 0 \\ 0 & \frac{m}{12}(L^2 + e^2) & 0 \\ 0 & 0 & m\frac{L^2}{6} \end{bmatrix} + m \begin{bmatrix} \frac{L^2}{2} + \frac{e^2}{2} & -\frac{L^2}{4} & -\frac{L.e}{4} \\ -\frac{L^2}{4} & \frac{L^2}{2} + \frac{e^2}{2} & -\frac{L.e}{4} \\ -\frac{L.e}{4} & -\frac{L.e}{4} & L^2 \end{bmatrix}$$

$$[I_O(S)] = m \begin{bmatrix} \frac{7}{12}(L^2 + e^2) & -\frac{L^2}{4} & -\frac{L.e}{4} \\ -\frac{L^2}{4} & \frac{7}{12}(L^2 + e^2) & -\frac{L.e}{4} \\ -\frac{L.e}{4} & -\frac{L.e}{4} & \frac{7}{6}L^2 \end{bmatrix}$$

4.3. Exercise 3: Determining the inertia matrix of a connecting rod

The connecting rod is a mechanical element that connects two moving joints in order to transmit a force. Figure 1 shows a model of a mechanically welded connecting rod made up of three parts. A first ring (**S1**) with outer radius **Re** and length **H** represents the connecting rod head. A second ring (**S2**) with outer radius re and length h represents the foot of the connecting rod, while a body (**S3**) is represented by a rectangular parallelepiped of length a, width b and thickness c. In this simplified model, we have omitted the junction between the body and the foot of the connecting rod, as well as between the body and the head of the connecting rod. All three parts are homogeneous and have a constant density ρ. The plane $(\vec{x}, \vec{y})$ is a plane of symmetry.

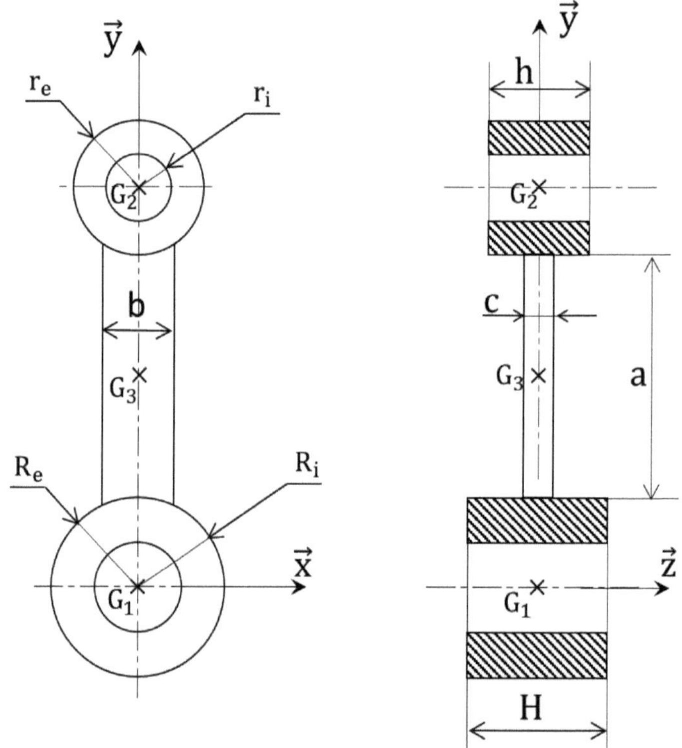

Figure 0.10 Model of a connecting rod

1) Determine, as a function of the geometric parameters of the problem and in the reference frame $(G_1, \vec{x}, \vec{y}, \vec{z})$, the position of the centre of mass of the connecting rod

2) In the reference frame $(G_1, \vec{x}, \vec{y}, \vec{z})$ the inertia matrix $[I_{(G_1, \vec{x}, \vec{y}, \vec{z})}(S1)]$ of the ring (S1).

3) In the reference frame $(G_2, \vec{x}, \vec{y}, \vec{z})$ the inertia matrix $[I_{(G_2, \vec{x}, \vec{y}, \vec{z})}(S2)]$ of the ring (S2).

4) In the reference frame $(G_3, \vec{x}, \vec{y}, \vec{z})$ the inertia matrix $[I_{(G_3, \vec{x}, \vec{y}, \vec{z})}(S3)]$ of the body (S3).

5) Apply Huygens' theorem to determine the matrices $[I_{(G_1, \vec{x}, \vec{y}, \vec{z})}(S2)]$ and $[I_{(G_1, \vec{x}, \vec{y}, \vec{z})}(S3)]$

6) In the reference frame $(G_1, \vec{x}, \vec{y}, \vec{z})$ the inertia matrix $[I_{(G_1, \vec{x}, \vec{y}, \vec{z})}(S1+S2+S3)]$ of the connecting rod.

Correction:

1 / The position of the centre of mass of the connecting rod in the reference frame $(G_1, \vec{x}, \vec{y}, \vec{z})$

For a discrete system, the centre of mass is calculated as follows:

$$m.\overrightarrow{G_1G} = \sum_{i=1}^{n} \overrightarrow{G_1G_i}.m_i$$

$$M.\overrightarrow{G_1G} = \sum_{i=1}^{3} \overrightarrow{G_1G_i}.m_i = m_1.\overrightarrow{G_1G_1} + m_2.\overrightarrow{G_1G_2} + m_3.\overrightarrow{G_1G_3} = m_2.\overrightarrow{G_1G_2} + m_3.\overrightarrow{G_1G_3}$$

Where M=m +m +m$_{123}$

$$M.\overrightarrow{G_1G} = m_2.\overrightarrow{G_1G_2} + m_3.\overrightarrow{G_1G_3} = m_2.\begin{pmatrix} 0 \\ R_e + a + r_e \\ 0 \end{pmatrix} + m_3.\begin{pmatrix} 0 \\ R_e + \frac{a}{2} \\ 0 \end{pmatrix}$$

2/ Ring inertia matrix (S1)

(S1) has two planes of symmetry $(o, \vec{y}, \vec{z})$ and $(o, \vec{x}, \vec{z})$ so D=E=F=0...

The axes $(o, \vec{x})$ and $(o, \vec{y})$ are equivalent, so $A = B$

$$[I_{(G_1, \vec{x}, \vec{y}, \vec{z})}(S1)] = \begin{bmatrix} A_e & 0 & 0 \\ 0 & A_e & 0 \\ 0 & 0 & C_e \end{bmatrix} - \begin{bmatrix} A_i & 0 & 0 \\ 0 & A_i & 0 \\ 0 & 0 & C_i \end{bmatrix}$$

$A = \int (x^2 + z^2).\,dm \quad (1)$

$A = \int (y^2 + z^2).\,dm \quad (2)$

$C = \int (x^2 + y^2).\,dm \quad (3)$

Let's add member to member the relations (1) and (2)

$$2. A = \int (x^2 + y^2). \, dm + 2. \int z^2. \, dm$$

This reveals C. from which

$$A = \frac{C}{2} + \int z^2. \, dm \text{ with } dm = \rho. r. dr. d\theta. dz$$

$$dm = \rho. \pi. R^2. dz = \frac{m}{h} dz \text{ hence } \int z^2. dm = \frac{m}{h}. \int z^2. dz = \frac{m.h^2}{12}$$

$$C = \int (x^2 + y^2). dm = \int r^2. dm = \frac{2.m}{R^2} \int r^3. dr = \frac{m.R^2}{2}$$

$$A = \frac{C}{2} + \int z^2. dm = \frac{m.R^2}{4} + \frac{m.h^2}{12}$$

Hence

$$[I_{(G_1,\vec{x},\vec{y},\vec{z})}(S1)] = m_{e1}. \begin{bmatrix} \frac{R_e^2}{4} + \frac{H^2}{12} & 0 & 0 \\ 0 & \frac{R_e^2}{4} + \frac{H^2}{12} & 0 \\ 0 & 0 & \frac{R_e^2}{4} \end{bmatrix} - m_{i1}. \begin{bmatrix} \frac{R_i^2}{4} + \frac{H^2}{12} & 0 & 0 \\ 0 & \frac{R_i^2}{4} + \frac{H^2}{12} & 0 \\ 0 & 0 & \frac{R_i^2}{4} \end{bmatrix}$$

With

Mass of the external volume $m_{e1} = \rho. \pi. R_e^2. H$

Mass of the interior volume $m_{i1} = \rho. \pi. R_i^2. H$

3/ Ring inertia matrix (S2)

(S2) has two planes of symmetry $(o, \vec{y}, \vec{z})$ and $(o, \vec{x}, \vec{z})$ so D=E=F=0.

The axes $(o, \vec{x})$ and $(o, \vec{y})$ are equivalent, so $A = B$

$$[I_{(G_2,\vec{x},\vec{y},\vec{z})}(S2)] = \begin{bmatrix} A_e & 0 & 0 \\ 0 & A_e & 0 \\ 0 & 0 & C_e \end{bmatrix} - \begin{bmatrix} A_i & 0 & 0 \\ 0 & A_i & 0 \\ 0 & 0 & C_i \end{bmatrix}$$

$$[I_{(G_2,\vec{x},\vec{y},\vec{z})}(S2)] = m_{e2} \begin{bmatrix} \frac{r_e^2}{4} + \frac{h^2}{12} & 0 & 0 \\ 0 & \frac{r_e^2}{4} + \frac{h^2}{12} & 0 \\ 0 & 0 & \frac{r_e^2}{4} \end{bmatrix} - m_{i2} \begin{bmatrix} \frac{r_i^2}{4} + \frac{h^2}{12} & 0 & 0 \\ 0 & \frac{r_i^2}{4} + \frac{h^2}{12} & 0 \\ 0 & 0 & \frac{r_i^2}{4} \end{bmatrix}$$

With

Mass of the external volume $m_{e2} = \rho. \pi. r_e^2. h$

Mass of the interior volume $m_{i2} = \rho. \pi. r_i^2. h$

4/ Matrix of inertia $[I_{(G_3,\vec{x},\vec{y},\vec{z})}(S3)]$ of body (S3) in $(G_3, \vec{x}, \vec{y}, \vec{z})$

(S3) has two planes of symmetry $(G_3, \vec{y}, \vec{z})$ and $(G_3, \vec{x}, \vec{z})$ so D=E=F=0.

$$[I_{(G_3,\vec{x},\vec{y},\vec{z})}(S3)] = \begin{bmatrix} A & 0 & 0 \\ 0 & B & 0 \\ 0 & 0 & C \end{bmatrix}$$

$$A = \int (y^2 + z^2).dm \qquad (1)$$

$$B = \int (x^2 + z^2).dm \qquad (2)$$

$$C = \int (x^2 + y^2).dm \qquad (3)$$

$$I_{(G_3,\vec{x},\vec{y},\vec{z})}(S3) = \begin{bmatrix} \dfrac{m_3}{12}(a^2 + c^2) & 0 & 0 \\ 0 & \dfrac{m_3}{12}(b^2 + c^2) & 0 \\ 0 & 0 & \dfrac{m_3}{12}(b^2 + a^2) \end{bmatrix}$$

6/ Apply Huygens' theorem to write the matrices $[I_{(G_1,\vec{x},\vec{y},\vec{z})}(S2)]$ and $[I_{(G_1,\vec{x},\vec{y},\vec{z})}(S3)]$

$$\left[I_{G_1}(S2)\right] = \left[I_{G_2}(S2)\right] + \left[I(G_2.G_1.m_2)\right]$$

$$\left[I_{G_2}(S2)\right] = m_{e2}\begin{bmatrix} \dfrac{r_e^2}{4} + \dfrac{h^2}{12} & 0 & 0 \\ 0 & \dfrac{r_e^2}{4} + \dfrac{h^2}{12} & 0 \\ 0 & 0 & \dfrac{r_e^2}{4} \end{bmatrix} - m_{i2}\begin{bmatrix} \dfrac{r_i^2}{4} + \dfrac{h^2}{12} & 0 & 0 \\ 0 & \dfrac{r_i^2}{4} + \dfrac{h^2}{12} & 0 \\ 0 & 0 & \dfrac{r_i^2}{4} \end{bmatrix}$$

$$[I(G_2.G_1.m_2)] = m_{e2}\begin{bmatrix} (R_e + a + r_e)^2 & 0 & 0 \\ 0 & 0 & 0 \\ 0 & 0 & (R_e + a + r_e)^2 \end{bmatrix} -$$

$$m_{i2}\begin{bmatrix} (R_i + a + r_i)^2 & 0 & 0 \\ 0 & 0 & 0 \\ 0 & 0 & (R_i + a + r_i)^2 \end{bmatrix}$$

$$\left[I_{G_1}(S3)\right] = \left[I_{G_3}(S3)\right] + \left[I(G_3.G_1.m_3)\right]$$

$$[I_{G1}(S3)] = \begin{bmatrix} \dfrac{m_3}{12}(a^2 + c^2) & 0 & 0 \\ 0 & \dfrac{m_3}{12}(b^2 + c^2) & 0 \\ 0 & 0 & \dfrac{m_3}{12}(b^2 + a^2) \end{bmatrix} + m_3\begin{bmatrix} \left(\dfrac{a}{2} + R_e\right)^2 & 0 & 0 \\ 0 & 0 & 0 \\ 0 & 0 & \left(\dfrac{a}{2} + R_e\right)^2 \end{bmatrix}$$

CHAPTER 2 : Kinetics

2. Kinetic torsor

2.1. Definition

The kinetic torsor, or torsor of the quantities of motion of a material system (S) in its motion relative to a reference frame R, is the torsor defined by :

$$\{C(S/R)\}_A = \left\{ \begin{array}{l} \vec{Q} = \int \vec{V}_{M/R}\, dm \\ \vec{\sigma_A}(S/R) = \int \overrightarrow{AM} \wedge \vec{V}_{M/R}dm \end{array} \right\} \tag{0.1}$$

Where

$$\vec{V}_{M/R} = \left[\frac{d\overrightarrow{OM}}{dt}\right]_R \tag{0.2}$$

$\vec{Q}$ is the kinetic resultant, also known as the momentum of (S). It is calculated from the velocity vector and depends on the reference frame in which it is calculated.

$\vec{\sigma}$: is the angular momentum at point A of (S) with respect to R this quantity depends on the reference frame in which it is calculated $\forall$ B.

$$\vec{\sigma_B}(S/R) = \int \overrightarrow{BM} \wedge \vec{V}_{M/R}dm = \int\left(\overrightarrow{BA} + \overrightarrow{AM}\right) \wedge \vec{V}_{M/R}dm \tag{0.3}$$

$$\vec{\sigma_B}(S/R) = \int \overrightarrow{BA} \wedge \vec{V}_{M/R}dm + \int \overrightarrow{AM} \wedge \vec{V}_{M/R}dm = \overrightarrow{BA} \wedge \int \vec{V}_{M/R}dm + \vec{\sigma_A}(S/R) \tag{0.4}$$

$$\vec{\sigma_B}(S/R) = \vec{\sigma_A}(S/R) + \overrightarrow{BA} \wedge \vec{Q} \tag{0.5}$$

2.2. Kinetic resultant

If G is the centre of gravity of (S) then

$$m.\overrightarrow{OG} = \int \overrightarrow{OM}.\,dm \tag{0.6}$$

Derived with respect to time :

$$\left[\frac{d(m.\overrightarrow{OG})}{dt}\right]_R = \left[\frac{d(\int \overrightarrow{OM}.dm)}{dt}\right]_R \tag{0.7}$$

Assuming that m is independent of time :

$$m\left[\frac{d\overrightarrow{OG}}{dt}\right]_R = \int \left[\frac{d\overrightarrow{OM}}{dt}\right]_R dm \tag{0.8}$$

So

$$m\,\vec{V}_{G/R} = \vec{Q} \tag{0.9}$$

$$\{C(S/R)\}_A = \left\{ \begin{array}{l} m\,\vec{V}_{G/R} \\ \vec{\sigma_A}(S/R) \end{array} \right\} \tag{0.10}$$

2.3. Kinetic moment

If (S) is a material system with centre of gravity G, the kinematic torsor $\{V_{(S/R)}\}$ of solid S at point A is :

$$\{V_{S/R}\}_A = \begin{Bmatrix} \vec{\Omega}_{S/R} \\ \vec{V}_{A\in S/R} \end{Bmatrix}_A \tag{0.11}$$

The angular momentum at point A is :

$$\vec{\sigma_A}(S/R) = \int \overrightarrow{AM} \wedge \vec{V}_{M\in S/R}\, dm \tag{0.12}$$

Now we can write :

$$\vec{V}_{M\in S/R} = \vec{V}_{A\in S/R} + \overrightarrow{MA} \wedge \vec{\Omega}_{S/R} \tag{0.13}$$

So

$$\vec{\sigma_A}(S/R) = \int \overrightarrow{AM} \wedge \vec{V}_{A\in S/R}\, dm + \int \overrightarrow{AM} \wedge (\overrightarrow{MA} \wedge \vec{\Omega}_{S/R})\, dm \tag{0.14}$$

$$\vec{\sigma_A}(S/R) = \int \overrightarrow{AM} \wedge \vec{V}_{A\in S/R}\, dm + \int \overrightarrow{AM} \wedge (\vec{\Omega}_{S/R} \wedge \overrightarrow{AM})\, dm \tag{0.15}$$

$$\vec{\sigma_A}(S/R) = m.\overrightarrow{AG} \wedge \vec{V}_{A\in S/R} + J_A(S)(\vec{\Omega}_{S/R}) \tag{0.16}$$

$$\vec{\sigma_A}(S/R) = m.\overrightarrow{AG} \wedge \vec{V}_{A\in S/R} + [I_A(s)].\vec{\Omega}_{S/R} \tag{0.17}$$

Note:

- If A coincides with the centre of gravity G, then

$$\vec{\sigma_G}(S/R) = [I_G(s)].\vec{\Omega}_{S/R} \tag{0.18}$$

- If A is fixed in R, then

$$\vec{\sigma_A}(S/R) = [I_A(s)].\vec{\Omega}_{S/R} \tag{0.19}$$

- If the speed of A is parallel $\overrightarrow{AG}$ then

$$\vec{\sigma_A}(S/R) = [I_A(s)].\vec{\Omega}_{S/R} \tag{0.20}$$

3. Dynamic torsor

2.1. Definition

The dynamic torsor, also known as the torsor of acceleration quantities, of a material system (S) in motion relative to R at any point A is :

$$\{D_{S/R}\}_A = \begin{Bmatrix} \int \vec{\Gamma}_{M/R}\, dm \\ \vec{\delta_A}(S/R) = \int \overrightarrow{AM} \wedge \vec{\Gamma}_{M/R}\, dm \end{Bmatrix} \tag{0.21}$$

- $\int \vec{\Gamma}_{M/R}\, dm$ is the dynamic resultant (also called the acceleration quantity) of (S).

- $\vec{\delta_A}(S/R)$ is the resultant moment or dynamic moment of (S) with respect to R.

2.2. Dynamic result

By deriving the expression of the kinetic resultant with the mass is constant

$$\left[\frac{d\, m.\vec{V}_{G/R}}{dt}\right]_R = \left[\frac{d \int \vec{V}_{M/R}\, dm}{dt}\right]_R \tag{0.22}$$

$$m\left[\frac{d\vec{V}_{G/R}}{dt}\right]_R = \int \vec{\Gamma}_{M/R}\, dm \tag{0.23}$$

$$\int \vec{\Gamma}_{M/R}\, dm = m.\,\vec{\Gamma}_{G/R} \tag{0.24}$$

2.3. Dynamic moment

The angular momentum at A of S relative to R is

$$\vec{\sigma_A}(S/R) = \int \overrightarrow{AM} \wedge \vec{V}_{M\in S/R}\, dm \tag{0.25}$$

Deriving this equation

$$\left[\frac{d}{dt}\vec{\sigma_A}(S/R)\right]_R = \int \left[\frac{d}{dt}\overrightarrow{AM} \wedge \vec{V}_{M/R}\right]_R dm \tag{0.26}$$

$$\left[\frac{d}{dt}\vec{\sigma_A}(S/R)\right]_R = \int \left[\frac{d}{dt}\overrightarrow{AM}\right]_R \wedge \vec{V}_{M/R}\, dm + \int \overrightarrow{AM} \wedge \left[\frac{d}{dt}\vec{V}_{M/R}\right]_R dm \tag{0.27}$$

Or

$$\overrightarrow{AM} = \overrightarrow{OM} - \overrightarrow{OA} \tag{0.28}$$

$$\left[\frac{d}{dt}\vec{\sigma_A}(S/R)\right]_R = \int \left[\frac{d\overrightarrow{OM}}{dt} - \frac{d\overrightarrow{OA}}{dt}\right]_R \wedge \vec{V}_{M/R}\, dm + \int \overrightarrow{AM} \wedge \left[\frac{d}{dt}\vec{V}_{M/R}\right]_R dm \tag{0.29}$$

$$\left[\frac{d}{dt}\vec{\sigma_A}(S/R)\right]_R = \int (\vec{V}_{M/R} - \vec{V}_{A/R}) \wedge \vec{V}_{M/R}\, dm + \int \overrightarrow{AM} \wedge \vec{\Gamma}_{M/R}\, dm \tag{0.30}$$

$$\left[\frac{d}{dt}\vec{\sigma_A}(S/R)\right]_R = \int \vec{V}_{M/R} \wedge \vec{V}_{M/R}\, dm - \int \vec{V}_{A/R} \wedge \vec{V}_{M/R}\, dm + \int \overrightarrow{AM} \wedge \vec{\Gamma}_{M/R}\, dm \tag{0.31}$$

$$\left[\frac{d}{dt}\vec{\sigma_A}(S/R)\right]_R = -\int \vec{V}_{A/R} \wedge \vec{V}_{M/R}\, dm + \int \overrightarrow{AM} \wedge \vec{\Gamma}_{M/R}\, dm \tag{0.32}$$

The dynamic moment $\int \overrightarrow{AM} \wedge \vec{\Gamma}_{M/R}\, dm$ noted $\vec{\delta_A}(S/R)$ can be formulated as follows :

$$\vec{\delta_A}(S/R) = \left[\frac{d}{dt}\vec{\sigma_A}(S/R)\right]_R + \int \vec{V}_{A/R} \wedge \vec{V}_{M/R}\, dm \tag{0.33}$$

$$\vec{\delta_A}(S/R) = \left[\frac{d}{dt}\vec{\sigma_A}(S/R)\right]_R + m.\,\vec{V}_{A/R} \wedge \vec{V}_{G/R} \tag{0.34}$$

<u>*Note:*</u>

- If A is the centre of gravity $\vec{\delta_G}(S/R) = \left[\frac{d}{dt}\vec{\sigma_A}(S/R)\right]_R$

- If A is fixed in R, the speed of A is parallel $\vec{V}_{G/R}$ $\quad \vec{\delta_A}(S/R) = \left[\frac{d}{dt}\vec{\sigma_A}(S/R)\right]_R$

<u>*Summary:*</u>

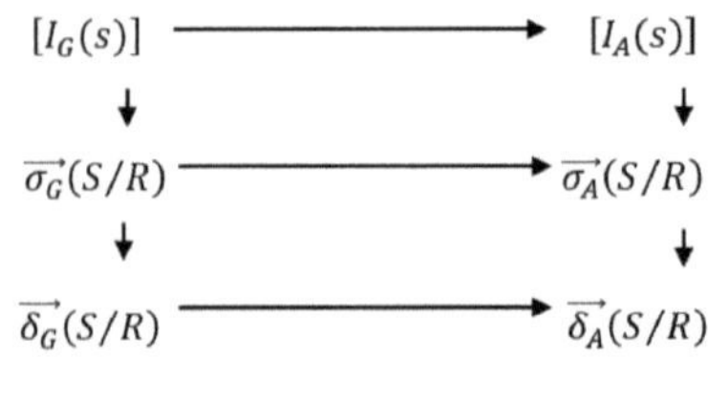

4. Application

4.1. Exercise 1: Kinetic and dynamic torsors of a rod

Consider a homogeneous rod of length l, negligible thickness and mass m in pivotal connection of axis $(O, \vec{z})$ with the frame and contained in the plane ($O, \vec{x}, \vec{y}$) .

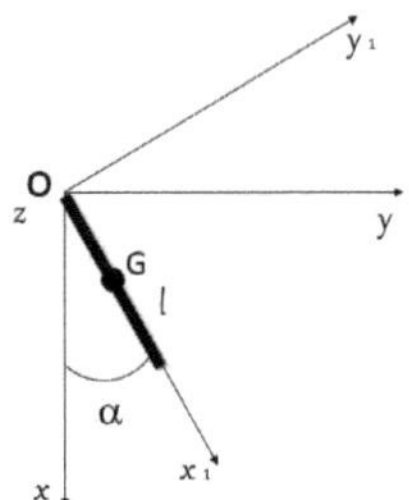

Figure 0.1 Simple pendulum

Kinetic torsor

Express the kinetic torsor at point G, the centre of mass, and point O, the origin of the reference frame.

$$\{C(T/R)\}_G = \left\{\begin{matrix} \vec{Q} \\ \vec{\sigma_G}(T/R) \end{matrix}\right\} = \left\{\begin{matrix} m\,\vec{V}_{G/R} \\ \vec{\sigma_G}(T/R) = \int \vec{GM} \wedge \vec{V}_{M/R}dm \end{matrix}\right\} \tag{0.35}$$

$$\int \vec{GM} \wedge \vec{V}_{M/R}dm = \int_0^l (x-\tfrac{1}{2})\vec{x_1} \wedge (x\dot{\alpha}\vec{y_1})dm = \int_0^l \left(x-\tfrac{1}{2}\right)\vec{x_1} \wedge (x\dot{\alpha}\vec{y_1})\tfrac{m}{l}dx \tag{0.36}$$

$$\{C(T/R)\}_G = \left\{\begin{matrix} \vec{Q} \\ \vec{\sigma_G}(T/R) \end{matrix}\right\} = \left\{\begin{matrix} \vec{Q} = m.\tfrac{1}{2}.\dot{\alpha}\,\vec{y_1} \\ \vec{\sigma_G}(T/R) = \tfrac{m.l^2}{12}\dot{\alpha}\,\vec{z} \end{matrix}\right\} \tag{0.37}$$

Dynamic torsor

Express the dynamic torsor of the rod at the centre of mass G.

$$\{C(T/R)\}_G = \left\{\begin{matrix} \vec{Q} \\ \vec{\sigma_G}(T/R) \end{matrix}\right\} = \left\{\begin{matrix} \vec{Q} = m.\tfrac{1}{2}.\dot{\alpha}\,\vec{y_1} \\ \vec{\sigma_G}(T/R) = \tfrac{m.l^2}{12}\dot{\alpha}\,\vec{z} \end{matrix}\right\} \tag{0.38}$$

$$\{D_{T/R}\}_G = \left\{\begin{matrix} m.\vec{\Gamma}_{G/R} \\ \vec{\delta_G}(T/R) = \left[\tfrac{d}{dt}\vec{\sigma_G}(S/R)\right]_R \end{matrix}\right\} \tag{0.39}$$

$$\vec{\Gamma}_{G/R} = \left[\tfrac{d}{dt}\left(\tfrac{1}{2}.\dot{\alpha}\,\vec{y_1}\right)\right]_R = \tfrac{1}{2}.\ddot{\alpha}.\vec{y_1} - \tfrac{1}{2}.\dot{\alpha}^2.\vec{x_1} \tag{0.40}$$

$$m.\vec{\Gamma}_{G/R} = \tfrac{m.l}{2}(\ddot{\alpha}.\vec{y_1} - \dot{\alpha}^2.\vec{x_1}) \tag{0.41}$$

$$\vec{\delta_G}(T/R) = \left[\tfrac{d}{dt}\vec{\sigma_G}(T/R)\right]_R = \left[\tfrac{d}{dt}\left(\tfrac{m.l^2}{12}\dot{\alpha}\,\vec{z}\right)\right]_R = \tfrac{m.l^2}{12}\ddot{\alpha}\,\vec{z} \tag{0.42}$$

$$\{D_{T/R}\}_G = \left\{ \begin{array}{c} m.\vec{\Gamma}_{G/R} \\ \vec{\delta_G}(T/R) = \left[\frac{d}{dt}\vec{\sigma_G}(T/R)\right]_R \end{array} \right\} = \left\{ \begin{array}{c} \frac{m.l}{2}(\ddot{\alpha}.\vec{y_1} - \dot{\alpha}^2.\vec{x_1}) \\ \vec{\delta_G}(T/R) = \frac{m.l^2}{12}\ddot{\alpha}\,\vec{z} \end{array} \right\} \tag{0.43}$$

$$\vec{\delta_0}(T/R) = \vec{\delta_G}(T/R) + \vec{OG} \wedge m.\vec{\Gamma}_{\underline{G}}_R = \frac{m.l^2}{12}\ddot{\alpha}\,\vec{z} + \frac{m.l^2}{4}\ddot{\alpha}\,\vec{z} = \frac{m.l^2}{3}\ddot{\alpha}\,\vec{z} \tag{0.44}$$

4.2. Exercise 2: Wheel on a horizontal plane

Consider the wheel (S) shown in Figure 2.2. It is modelled by a homogeneous disc of radius R, negligible thickness and mass m. The position of the centre of mass G relative to a reference frame R assumed to be fixed is denoted x. The point of contact with the ground is denoted I. Determine, relative to the reference frame R, the kinetic $\{C_{(S/R)}\}_G$ and dynamic $\{D_{(S/R)}\}_G$ at point G.

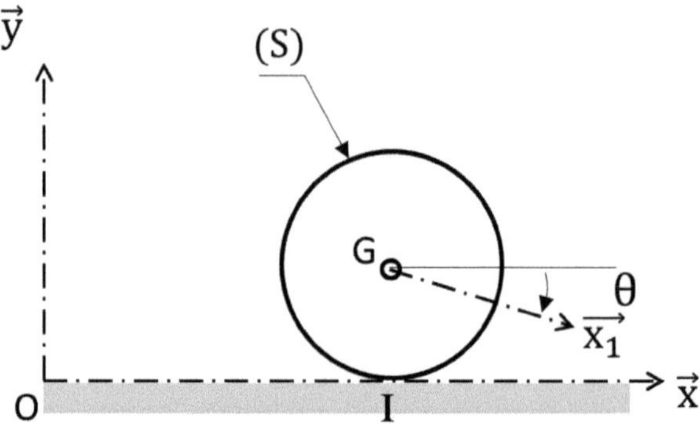

Figure 0.2 Wheel on horizontal plane

Calculating the kinetic resultant. By definition :

$$\vec{Q} = m\,\vec{V}_{G/R} = m\frac{d\vec{OG}}{dt}\bigg|_R = m\frac{d(R.\theta.\vec{x} + R.\vec{y})}{dt}\bigg|_R = m.R.\dot{\theta}.\vec{x}$$

Calculation of angular momentum at G. By definition :

$$\vec{\sigma_G}(S/R) = [I_G(s)].\vec{\Omega}_{S/R}$$

$$\vec{\sigma_G}(S/R) = \begin{bmatrix} m\frac{R^2}{4} & 0 & 0 \\ 0 & m\frac{R^2}{4} & 0 \\ 0 & 0 & m\frac{R^2}{2} \end{bmatrix} . \begin{pmatrix} 0 \\ 0 \\ \dot{\theta} \end{pmatrix} = m\frac{R^2}{2}\dot{\theta}\,\vec{z}$$

$$\{C_{(S/R)}\}_G = \left\{ \begin{array}{c} \vec{Q} = m.R.\dot{\theta}.\vec{x} \\ \vec{\sigma_G}(S/R) = m\frac{R^2}{2}\dot{\theta}\,\vec{z} \end{array} \right\}_G$$

By deriving the expression of the kinetic resultant with the mass is constant

$$\int \vec{r}_{M/R}\,dm = m.\vec{r}_{G/R}$$

$$m.\vec{r}_{G/R} = m.R.\ddot{\theta}.\vec{x}$$

Dynamic moment

$$\vec{\delta_G}(S/R) = \left[\frac{d}{dt}\vec{\sigma_G}(S/R)\right]_R = m\frac{R^2}{2}\ddot{\theta}\,\vec{z}$$

$$\{D_{(S/R)}\}_G = \begin{Bmatrix} m.\vec{r}_{G/R} = m.R.\ddot{\theta}.\vec{x} \\ \vec{\delta_G}(S/R) = m\frac{R^2}{2}\ddot{\theta}\,\vec{z} \end{Bmatrix}_G$$

4.3. Exercise 3: Merry-go-round

The device shown (Figure 2.3) models the movement of a merry-go-round: it consists of a bar (S1) of length $OA = l$ whose mass is neglected in relation to the rest of the device - connected to a solid (S2) which represents the gondola in which the people are installed; this gondola is modelled by a homogeneous disc of radius R and mass m; its thickness is assumed to be negligible. The orientation of the bar (S1) relative to the reference frame R $(O, \vec{x_0}, \vec{y_0}, \vec{z})$ linked to the ground (S0) is given by the angle θ_1 and that of the gondola (S2) relative to the solid (S1) by the angle θ_2. The distance $AG = a$.

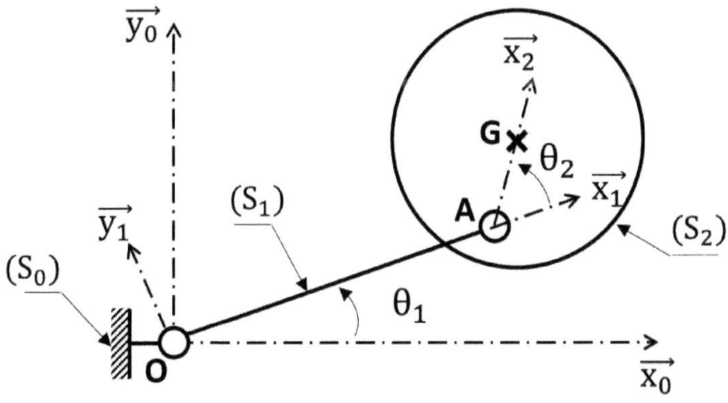

Figure 0.3 Carousel

1- Determine, relative to the reference frame R, the kinetic torsor at point A of the assembly (S1+S2)

2- Determine, relative to the reference frame R, the dynamic torsor at point A of the assembly (S1+S2)

Matrix of inertia of a rod at point G with axis (ox)

$$[I_G(S)] = \begin{bmatrix} 0 & 0 & 0 \\ 0 & m\dfrac{l^2}{3} & 0 \\ 0 & 0 & m\dfrac{l^2}{3} \end{bmatrix}$$

Matrix of inertia of a disc at point G of axis (oz)

$$[I_G(S)] = \begin{bmatrix} m\dfrac{R^2}{4} & 0 & 0 \\ 0 & m\dfrac{R^2}{4} & 0 \\ 0 & 0 & m\dfrac{R^2}{2} \end{bmatrix}$$

Correction:

1/ The kinetic torsor of the assembly is (by definition) the sum of the kinetic torsorsors of each of the elements. Then, as S1 is considered to have no mass (or its mass is considered to have no dynamic effect), its kinetic torsor is zero.

We begin by writing the kinematic torsor of S2/S0 :

$$\{V_{(S2/R)}\}_G = \left\{ \begin{array}{c} \vec{\Omega}_{(S2/S0)} \\ \vec{V}_{(G\in S2/S0)} \end{array} \right\}_G$$

For the rotation vector :

$$\vec{\Omega}_{(S2/S0)} = \left(\dot{\theta}_1 + \dot{\theta}_2\right)\vec{z}$$

For the moment (speed) in G :

$$\vec{V}_{(G\in S2/S0)} = \left.\frac{d\overrightarrow{OG}}{dt}\right|_R = \left.\frac{d(l.\vec{x_1}+a.\vec{x_2})}{dt}\right|_R = l.\dot{\theta}_1.\vec{y_1} + a.\left(\dot{\theta}_1 + \dot{\theta}_2\right).\vec{y_2}$$

Calculation of the kinetic resultant of S2/S0. By definition :

$$\vec{Q} = m.\vec{V}_{(G\in S2/S0)}$$

Calculation of angular momentum at G. By definition :

$$\vec{\sigma_G}(S2/S0) = [I_G(S2)].\vec{\Omega}_{\,S2/R}$$

Calculation of angular momentum at A. By definition :

$$\vec{\sigma_A}(S2/S0) = m.\overrightarrow{AG} \wedge \vec{V}_{(A\in S2/S0)} + [I_A(S2)].\vec{\Omega}_{\,S2/R}$$

However, the inertia matrix is easier to express at the centre of gravity G. There are therefore two equivalent solutions: the first is to calculate the matrix in G and then in A (using Huygens' theorem) and apply the formula above; the second is to calculate the angular momentum in G and deduce that in A using the point change formula:

$$\vec{\sigma_A}(S2/S0) = \vec{\sigma_G}(S2/S0) + m.\overrightarrow{AG} \wedge \vec{V}_{(G\in S2/S0)}$$

There are :

$$\vec{\sigma_G}(S2/S0) = \frac{m.R^2}{2}\left(\dot{\theta}_1 + \dot{\theta}_2\right)\vec{z}$$

Then deduce the angular momentum at point A using the point change formula:

$$\vec{\sigma_A}(S2/S0) = \frac{m.R^2}{2}\left(\dot{\theta}_1 + \dot{\theta}_2\right)\vec{z} + m.a.\overrightarrow{x_2} \wedge \left[l.\dot{\theta}_1.\overrightarrow{y_1} + a.\left(\dot{\theta}_1 + \dot{\theta}_2\right).\overrightarrow{y_2}\right]$$

The vector product expansion gives :

$$\vec{\sigma_A}(S2/S0) = \frac{m.R^2}{2}\left(\dot{\theta}_1 + \dot{\theta}_2\right)\vec{z} + m.a.l.\dot{\theta}_1.\overrightarrow{x_2} \wedge \overrightarrow{y_1} + m.a^2.\left(\dot{\theta}_1 + \dot{\theta}_2\right).\overrightarrow{x_2} \wedge \overrightarrow{y_2}$$

The projection of $\overrightarrow{x_2}$ in the base $(\overrightarrow{x_1}, \overrightarrow{y_1}, \vec{z})$

$$\vec{\sigma_A}(S2/S0) = \frac{m.R^2}{2}\left(\dot{\theta}_1 + \dot{\theta}_2\right)\vec{z} + m.a.l.\dot{\theta}_1.\left(\cos\theta_2.\overrightarrow{x_1} + \sin\theta_2.\overrightarrow{y_1}\right) \wedge \overrightarrow{y_1} + m.a^2.\left(\dot{\theta}_1 + \dot{\theta}_2\right).\vec{z}$$

And, in the end :

$$\vec{\sigma_A}(S2/S0) = \left[m\left(\frac{R^2}{2} + a^2\right)\left(\dot{\theta}_1 + \dot{\theta}_2\right) + m.a.l.\dot{\theta}_1.\cos\theta_2\right]\vec{z}$$

2/ The dynamic torsor of the assembly is (by definition) the sum of the dynamic torsorsors of each of the elements. Then, as S1 is considered to have no mass (or its mass is considered to have no dynamic effect), its dynamic torsor is zero.

Calculation of the dynamic resultant of S2/S0. By definition :

$$m.\vec{\Gamma}_{(G\in S2/S0)} = m.\left.\frac{d\vec{V}_{(G\in S2/S0)}}{dt}\right|_R$$

We can see that we need to derive two moving vectors in the 0 basis. We apply the derivation formula in a moving basis. We find :

$$m.\vec{\Gamma}_{(G\in S2/S0)} = m.\left.\frac{d\left(1.\dot{\theta}_1.\vec{y_1} + a.(\dot{\theta}_1 + \dot{\theta}_2).\vec{y_2}\right)}{dt}\right|_R$$

Calculating the derivative gives :

$$m.\vec{\Gamma}_{(G\in S2/S0)} = m.\left[1.\ddot{\theta}_1.\vec{y_1} - 1.\dot{\theta}_1^{\,2}.\vec{x_1} + a.(\ddot{\theta}_1 + \ddot{\theta}_2).\vec{y_2} - a.(\dot{\theta}_1 + \dot{\theta}_2)^2.\vec{x_2}\right]$$

Calculation of the dynamic moment at G: two solutions are possible:

the dynamic moment at A is calculated directly from the general formula :

$$\vec{\delta_A}(S2/S0) = \left[\frac{d}{dt}\vec{\sigma_A}(S2/S0)\right]_0 + m.\vec{V}_{(A\in S2/S0)} \wedge \vec{V}_{(G\in S2/S0)}$$

the dynamic moment at G is calculated from the angular momentum at G and then calculated at A using the point change formula

$$\vec{\delta_A}(S2/S0) = \vec{\delta_G}(S2/S0) + \vec{AG} \wedge m.\vec{\Gamma}_{(G\in S2/S0)}$$

The dynamic moment at G is

$$\vec{\delta_G}(S2/S0) = \left[\frac{d}{dt}\vec{\sigma_G}(S2/S0)\right]_R = \left[\frac{d}{dt}\left(\frac{m.R^2}{2}(\dot{\theta}_1 + \dot{\theta}_2)\,\vec{z}\right)\right]_R = \frac{m.R^2}{2}(\ddot{\theta}_1 + \ddot{\theta}_2)\,\vec{z}$$

The second term of the dynamic moment expression can be calculated as follows

$$\vec{AG} \wedge m.\vec{\Gamma}_{(G\in S2/S0)} = m.a.\vec{x_2} \wedge \left[1.\ddot{\theta}_1.\vec{y_1} - 1.\dot{\theta}_1^{\,2}.\vec{x_1} + a.(\ddot{\theta}_1 + \ddot{\theta}_2).\vec{y_2} - a.(\dot{\theta}_1 + \dot{\theta}_2)^2.\vec{x_2}\right]$$

$$\vec{AG} \wedge m.\vec{\Gamma}_{(G\in S2/S0)}$$

$$= m.a.\vec{x_2} \wedge \left[1.\ddot{\theta}_1.\vec{y_1} - 1.\dot{\theta}_1^{\,2}.\vec{x_1}\right] + m.a.\vec{x_2}$$

$$\wedge \left[a.(\ddot{\theta}_1 + \ddot{\theta}_2).\vec{y_2} - a.(\dot{\theta}_1 + \dot{\theta}_2)^2.\vec{x_2}\right]$$

The projection of $\vec{x_2}$ in the base $(\vec{x_1}, \vec{y_1}, \vec{z})$

$$\vec{AG} \wedge m.\vec{\Gamma}_{(G\in S2/S0)} = m.a.(\cos\theta_2.\vec{x_1} + \sin\theta_2.\vec{y_1}) \wedge \left[1.\ddot{\theta}_1.\vec{y_1} - 1.\dot{\theta}_1^{\,2}.\vec{x_1}\right] + m.a^2.(\ddot{\theta}_1 + \ddot{\theta}_2).\vec{z}$$

The calculation gives :

$$\vec{AG} \wedge m.\vec{\Gamma}_{(G\in S2/S0)} = m.a.\left[\cos\theta_2.1.\ddot{\theta}_1 + \sin\theta_2.1.\dot{\theta}_1^{\,2} + a.(\ddot{\theta}_1 + \ddot{\theta}_2)\right].\vec{z}$$

Finally, the expression for the dynamic moment at point A is

$$\overrightarrow{\delta_A}(S2/S0) = \left(\frac{m.R^2}{2}\left(\ddot{\theta}_1 + \ddot{\theta}_2\right) + m.a.\left[\cos\theta_2 .l.\ddot{\theta}_1 . + \sin\theta_2 .l.\dot{\theta}_1^{\,2} + a.\left(\ddot{\theta}_1 + \ddot{\theta}_2\right)\right]\right).\vec{z}$$

CHAPTER 3 : Fundamental principle of PFD dynamics

3. Fundamental principle

3.1. Statement

The Fundamental Principle of Dynamics (FPD) states that if a material system (Σ) in motion relative to a Galilean reference frame $\mathbf{R_G}$, then the dynamic torsor of (Σ) is equal to the torsor of the external mechanical actions.

If we denote the outside of (Σ) by ($\overline{\Sigma}$) then the PFD is written :

$$\{D_{\Sigma/Rg}\}_A = \{T_{\overline{\Sigma}->\Sigma}\}_A \forall \quad A \tag{0.1}$$

3.2. General theorems of dynamics

Consider a material system (Σ) of mass m and centre of mass G in motion relative to a reference frame $\mathbf{Rg}$, at any point A.

$$\{D_{\Sigma/Rg}\}_A = \{T_{\overline{\Sigma}->\Sigma}\}_A \tag{0.2}$$

$$\begin{Bmatrix} m.\vec{\Gamma}_{G/R} \\ \vec{\delta}_{A\in\Sigma/R} \end{Bmatrix}_A = \begin{Bmatrix} \vec{R}_{(\overline{\Sigma}->\Sigma)} \\ \vec{M}_{(\overline{\Sigma}->\Sigma)} \end{Bmatrix}_A \tag{0.3}$$

This leads to two vector equations known as general theorems of dynamics.

3.2.1. Dynamic resultant theorem

The resultant of the torsor of mechanical actions external to (Σ) is equivalent to the resultant of the dynamic torsor of ( ).Σ

$$\vec{R}_{(\overline{\Sigma}->\Sigma)} = m.\vec{\Gamma}_{G/R} \tag{0.4}$$

3.2.2. Dynamic moment theorem

The moment of the torsor of the mechanical actions external to (Σ) at any point A is equivalent to the moment of the dynamic torsor of (Σ)taken at the same point.

$$\forall A \; \vec{M}_{(\overline{\Sigma}->\Sigma)} = \vec{\delta}_{A\in\Sigma/R} \tag{0.5}$$

3.2.3. Theorem of mutual actions

Consider a hardware system (Σ) made up of two subsystems (Σ_1) and (Σ_2). The PFD is applied to the two subsystems (Σ_1) and (Σ_2) of (Σ). We find :

$$\{D_{\Sigma_1/Rg}\}_A + \{D_{\Sigma_2/Rg}\}_A = \{D_{\Sigma/Rg}\}_A \tag{0.6}$$

In addition, the PFD is applied to each sub-system.

$$\{D_{\Sigma_1/Rg}\}_A = \{T_{\overline{\Sigma_1}->\Sigma_1}\}_A = \{T_{\overline{\Sigma}->\Sigma_1}\}_A + \{T_{\Sigma_2->\Sigma_1}\}_A \tag{0.7}$$

$$\{D_{\Sigma_2/Rg}\}_A = \{T_{\overline{\Sigma_2}->\Sigma_2}\}_A = \{T_{\overline{\Sigma}->\Sigma_2}\}_A + \{T_{\Sigma_1->\Sigma_2}\}_A \tag{0.8}$$

Applying the PFD to the set (Σ) we find :

$$\{D_{\Sigma/Rg}\}_A = \{T_{\overline{\Sigma}->\Sigma}\}_A = \{T_{\overline{\Sigma}->\Sigma_1}\}_A + \{T_{\overline{\Sigma}->\Sigma_2}\}_A \tag{0.9}$$

By adding these two equations (3.7) and (3.8) and using equation (3.9) we find :

$$\{T_{\Sigma_1->\Sigma_2}\}_A = -\{T_{\Sigma_2->\Sigma_1}\}_A \tag{0.10}$$

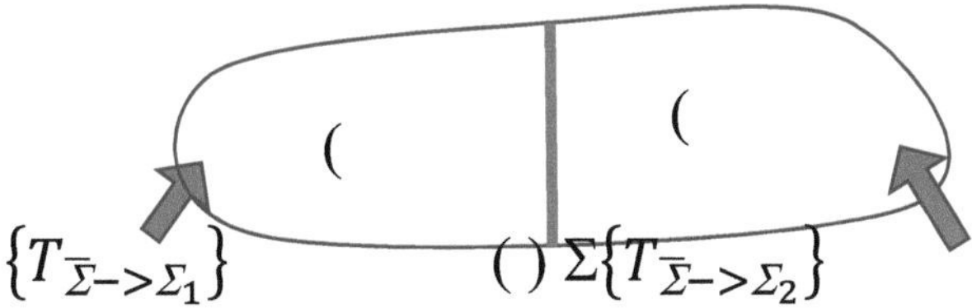

Figure 0.1 Subdivision of the system (Σ) into two parts (Σ 1) and (Σ 2).

3.2.4. Special case of statics

In the case where the dynamic torsor is zero, the PFD is simplified to the Fundamental Principle of Statics. This case arises in particular under the following conditions:

- When we neglect the effects of inertia on the material system (Σ).
- When the mass of the material system (Σ) is considered to be zero, for example, in the case of a perfect spring.
- When the material system (Σ) moves in uniform rectilinear translation with respect to the Galilean reference frame **Rg**.

3.3. Expression of the PFD in a non-galilean frame of reference

Consider **Rg** as a Galilean frame of reference and **R** as a frame of reference in any known motion relative to **Rg**. If we take a material subset (Σ_1) of the material system (Σ) in motion relative to **R**.

Consider a material subset (Σ_1) of the material system (Σ) in motion relative to a reference frame **R**, which is in motion relative to a Galilean reference frame **Rg**. The PFD applied to ($\sum_1$) with respect to **Rg** is expressed as :

$$\{D_{\Sigma_1/Rg}\}_A = \{T_{\overline{\Sigma_1}->\Sigma_1}\}_A \tag{0.11}$$

The reduction elements of the dynamic torsor at any point A are :

$$\{D_{\Sigma_1/Rg}\}_A = \left\{ \begin{array}{c} \int \vec{\Gamma}_{M/Rg}\, dm \\ \int \overrightarrow{AM} \wedge \vec{\Gamma}_{M/Rg} dm \end{array} \right\}_A \tag{0.12}$$

If **R** has an entraining motion relative to **Rg** which leads to the law of composition of accelerations. Using the composition of acceleration vectors, at point M between **R** and **Rg** , we find :

$$\vec{\Gamma}(M/R_g) = \vec{\Gamma}(M/R) + \vec{\Gamma}(M \in R/R_g) + 2.\vec{\Omega}(R/R_g) \wedge \vec{V}(M/R) \tag{0.13}$$

Introducing equation (3.13) into (3.12) we find :

$$\{D_{\Sigma_1/Rg}\}_A = \left\{ \begin{array}{l} \int [\vec{\Gamma}(M/R) + \vec{\Gamma}(M \in R/R_g) + 2.\vec{\Omega}(R/R_g) \wedge \vec{V}(M/R)]\, dm \\ \int \overrightarrow{AM} \wedge [\vec{\Gamma}(M/R) + \vec{\Gamma}(M \in R/R_g) + 2.\vec{\Omega}(R/R_g) \wedge \vec{V}(M/R)]dm \end{array} \right\}$$

$$(0.14)$$

Considering equation (3.14), the dynamic torsor $\{D_{\Sigma_1/Rg}\}_A$ can be decomposed into three torsors :

$$\{D_{\Sigma_1/Rg}\}_A = \{D_{\Sigma_1/R}\}_A - \{D_{ie(\Sigma_1,R/Rg)}\}_A - \{D_{ic(\Sigma_1,R/Rg)}\}_A \tag{0.15}$$

- the dynamic torsor of (Σ_1) in its movement relative to **R**:

$$\{D_{\Sigma_1/R}\}_A = \left\{ \begin{array}{l} \int \vec{\Gamma}(M/R)\, dm \\ \int \overrightarrow{AM} \wedge \vec{\Gamma}(M/R)dm \end{array} \right\} \tag{0.16}$$

- the torsor of the driving inertia effects on (Σ_1) in its movement relative to **R** and **Rg** :

$$\{D_{ie(\Sigma_1,R/Rg)}\}_A = \left\{ \begin{array}{l} - \int \vec{\Gamma}(M \in R/R_g)\, dm \\ - \int \overrightarrow{AM} \wedge \vec{\Gamma}(M \in R/R_g)dm \end{array} \right\} \tag{0.17}$$

- the Coriolis torsor of inertial effects on (Σ_1) as it moves relative to **R** and **Rg** :

$$\{D_{ic(\Sigma_1,R/Rg)}\}_A = \left\{ \begin{array}{l} - \int 2.\vec{\Omega}(R/R_g) \wedge \vec{V}(M/R)\, dm \\ - \int \overrightarrow{AM} \wedge 2.\vec{\Omega}(R/R_g) \wedge \vec{V}(M/R)dm \end{array} \right\} \tag{0.18}$$

Given equation (3.15), the PFD is written in the **R** frame of reference as :

$$\{D_{\Sigma_1/R}\}_A = \{D_{\Sigma_1/Rg}\}_A + \{D_{ie(\Sigma_1,R/Rg)}\}_A + \{D_{ic(\Sigma_1,R/Rg)}\}_A \tag{0.19}$$

The PFD can therefore be applied in any reference frame, by adding to the torsor of the external mechanical actions the torsor of the driving inertia effects and the torsor of the Coriolis inertia effects.

If the motion between frame **R** and frame **Rg** is a uniform rectilinear translation, then $\vec{\Gamma}(M \in R/R_g) = \vec{0}$ and $\vec{\Omega}(R/R_g) = \vec{0}.$

As a result, the two torsors of the driving inertia effects and the Coriolis inertia effects are zero, and the PFD is then written in the reference frame R :

$$\{D_{\Sigma_1/R}\}_A = \{T_{\overline{\Sigma_1} \to \Sigma_1}\}_A \tag{0.20}$$

This leads to the conclusion that any reference frame **R** which is in uniform rectilinear translation with respect to **Rg** is also Galilean.

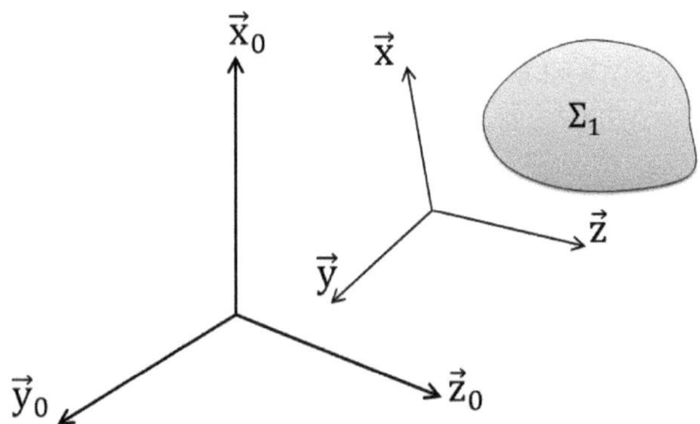

Figure 0.2 PFD in a non-galilean reference frame

4. Dynamic balancing :

One of the major challenges in the manufacture of rotating parts is dynamic balancing around an axis. This prevents the occurrence of mechanical vibrations, which can quickly damage bearings (causing premature fatigue) or simply hinder the use of the equipment.

4.1. Modelling

Consider a frame (S$_0$) associated with the Galilean reference frame $R_0(O, \vec{x_0}, \vec{y_0}, \vec{z_0})$. A solid (S$_1$), having a mass **m$_1$** and a centre of inertia G, is in a frictionless pivot connection of axis $(O, \vec{z_0})$ with (S$_0$).

Let $R(O, \vec{x}, \vec{y}, \vec{z_0})$ be a reference frame related to (S$_1$) chosen, to simplify calculations, so that the plane $(O, \vec{x}, \vec{z_0})$ contains the point G.

We pose

$(\vec{x_0}, \vec{x}) = \theta$ and $\overrightarrow{OG} = a.\vec{x} + c.\vec{z_0}$

Since the solid (S$_1$) is arbitrary, the inertia matrix of (S$_1$) at point O, in the base of R, is of the form :

$$[I_0(S_1)] = \begin{bmatrix} A & -F & -E \\ -F & B & -D \\ -E & -D & C \end{bmatrix}_{(\vec{x},\vec{y},\vec{z_0})} \tag{0.21}$$

The (unknown) mechanical action of (S_0) on (S_1) is modelled by the torsor $\{\tau_{(S_0 \to S_1)}\}$ at point O :

$$\{\tau_{(S_0 \to S_1)}\}_O = \left\{ \frac{\vec{R}}{\vec{M_O}} \right\} \tag{0.22}$$

Let's say :

$$\begin{cases} \vec{R} = X.\vec{x} + Y.\vec{y} + Z.\vec{z_0} \\ \vec{M_O} = L.\vec{x} + M.\vec{y} \end{cases} \tag{0.23}$$

Let $\{\tau_{(E \to S_1)}\}_O$ be the mechanical action exerted by a set of materials (E) on (S_1) at point O :

$$\{\tau_{(E \to S_1)}\}_O = \left\{ \frac{\vec{R_1}}{\vec{M_{1O}}} \right\} \tag{0.24}$$

Let's say :

$$\begin{cases} \vec{R_1} = X_1.\vec{x} + Y_1.\vec{y} + Z_1.\vec{z_0} \\ \vec{M_{1O}} = L_1.\vec{x} + M_1.\vec{y} + N_1.\vec{z_0} \end{cases} \tag{0.25}$$

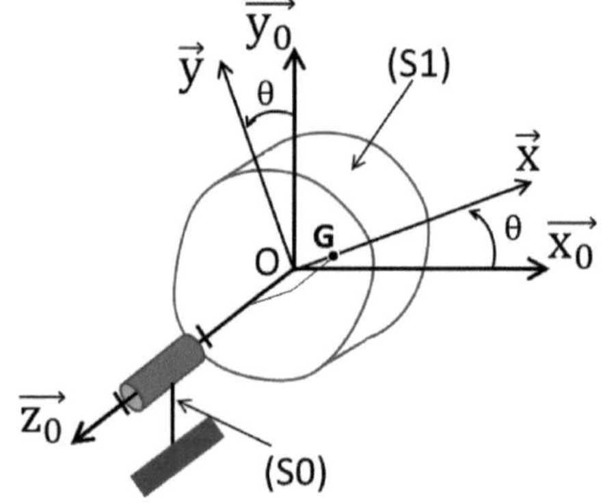

Figure 0.3 Modelling of the parameters of a solid rotating about $(O, \vec{Z})$

4.2. Mechanical actions

By applying the PFD to (S_1) moving relative to R_0 , the mechanical action torsor exerted by (S_0) on (S_1) can be determined.

$$\{D_{S_1/R_0}\}_O = \{T_{\overline{S_1} -> S_1}\}_O \tag{0.26}$$

Knowing that $(\overline{S_1})$ is made up of (S_0) and (E), this equality can be written as :

$$\{D_{S_1/R_0}\}_O = \{T_{S_0 -> S_1}\}_O + \{T_{E -> S_1}\}_O \tag{0.27}$$

The expression of these torsors at point O gives :

$$\left\{ \begin{matrix} m_1.\vec{\Gamma}_{G/R_0} \\ \vec{\delta_0}(S_1/R_0) \end{matrix} \right\} = \left\{ \begin{matrix} \vec{R} \\ \overrightarrow{M_O} \end{matrix} \right\} + \left\{ \begin{matrix} \overrightarrow{R_1} \\ \overrightarrow{M_{1_O}} \end{matrix} \right\} \tag{0.28}$$

The equation gives two vector equations:

$$\left\{ \begin{matrix} m_1.\vec{\Gamma}_{G/R_0} = \vec{R} + \overrightarrow{R_1} \\ \vec{\delta_0}(S_1/R_0) = \overrightarrow{M_O} + \overrightarrow{M_{1_O}} \end{matrix} \right. \tag{0.29}$$

The acceleration vector $\vec{\Gamma}_{G/R_0}$ is obtained by :

$$\vec{\Gamma}_{G/R_0} = \left[\frac{d}{dt} \vec{V}_{G/R_0} \right]_{R_0} \tag{0.30}$$

With

$$\vec{V}_{G/R_0} = a.\dot{\theta}.\vec{y} \tag{0.31}$$

So

$$\vec{\Gamma}_{G/R_0} = a.\ddot{\theta}.\vec{y} - a.\dot{\theta}^2.\vec{z_0} \tag{0.32}$$

Since the point O is fixed in R0, the dynamic moment $\vec{\delta_0}(S_1/R_0)$ is calculated from the angular momentum $\vec{\sigma_0}(S_1/R_0)$ by the relation :

$$\vec{\delta_0}(S_1/R_0) = \left[\frac{d}{dt} \vec{\sigma_0}(S_1/R_0) \right]_{R_0} \tag{0.33}$$

The angular momentum $\vec{\sigma_0}(S_1/R_0)$ is expressed as a function of the inertia operator by the relation :

$$\vec{\sigma_0}(S_1/R_0) = J_0(S_1)\left(\vec{\Omega}(S_1/R_0) \right) \tag{0.34}$$

With $\vec{\Omega}(S_1/R_0) = \dot{\theta}.\vec{z_0}$

This corresponds to the following matrix multiplication :

$$[\vec{\sigma_0}(S_1/R_0)] = \begin{bmatrix} A & -F & -E \\ -F & B & -D \\ -E & -D & C \end{bmatrix} . \begin{bmatrix} 0 \\ 0 \\ \dot{\theta} \end{bmatrix}_{(\vec{x},\vec{y},\vec{z_0})} \tag{0.35}$$

Hence

$$\vec{\sigma_0}(S_1/R_0) = -E.\dot{\theta}.\vec{x} - D.\dot{\theta}.\vec{y} + C.\dot{\theta}.\vec{z_0} \tag{0.36}$$

To calculate the dynamic moment $\vec{\delta_0}(S_1/R_0)$ from relation (3.33), let's use the derivation basis of the R frame.

$$\vec{\delta_0}(S_1/R_0) = \left[\frac{d}{dt} \vec{\sigma_0}(S_1/R_0) \right]_R + \vec{\Omega}(R/R_0) \wedge \vec{\sigma_0}(S_1/R_0) \tag{0.37}$$

Hence, with the expression for $\vec{\sigma_0}(S_1/R_0)$ from relation (3.36) :

$$\vec{\delta_0}(S_1/R_0) = -E.\ddot{\theta}.\vec{x} - D.\ddot{\theta}.\vec{y} + C.\ddot{\theta}.\vec{z_0} + \dot{\theta}.\vec{z_0} \wedge \left(-E.\dot{\theta}.\vec{x} - D.\dot{\theta}.\vec{y} + C.\dot{\theta}.\vec{z_0} \right)$$

$$\tag{0.38}$$

Or

$$\vec{\delta_0}(S_1/R_0) = \left(-E.\ddot{\theta} + D.\dot{\theta}^2 \right).\vec{x} - \left(D.\ddot{\theta} + E.\dot{\theta}^2 \right).\vec{y} + C.\ddot{\theta}.\vec{z_0} \tag{0.39}$$

The vector equation (3.29) deduced from the PFD is written in projection on the basis of R :

$$\begin{cases} \vec{x} & : \quad -m_1.a.\dot{\theta}^2 = X + X_1 \\ \vec{y} & : \quad -m_1.a.\ddot{\theta} = Y + Y_1 \\ \vec{z_0} & : \qquad 0 = Z + Z_1 \end{cases} \qquad (0.40)$$

$$\begin{cases} \vec{x} & : \quad -E.\ddot{\theta} + D.\dot{\theta}^2 = L + L_1 \\ \vec{y} & : \quad D.\ddot{\theta} + E.\dot{\theta}^2 = M + M_1 \\ \vec{z_0} & : \qquad C.\ddot{\theta} = N_1 \end{cases} \qquad (0.41)$$

From the equations we can easily express X, Y, Z, L and M

4.3. Dynamic balancing conditions

In order to minimise vibrations, the mechanical action of the link between (S_1) and (S_0) must be made independent of the movement of (S_1) relative to (S_0), i.e. of $\dot{\theta}$ and $\ddot{\theta}$.

According to equations (3.40) and (3.41) the dynamic balancing conditions are as follows:

- ✓ a = 0 the centre of inertia G is on the axis of rotation $(O, \vec{z_0})$ (static balancing)
- ✓ D = 0 et E = 0 the axis of rotation $(O, \vec{z_0})$ is a principal axis of inertia for (S_1).

4.4. Practical implementation of dynamic balancing

(S_1) is replaced by a solid (S') composed of (S_1) and two solids (S_2) and (S_3), considered as material points, so that the solid (S') is in dynamic equilibrium.

Let mi be the mass of the solid (Si) (i=2 and 3) placed at a point Mi (xi, yi, zi) of the reference frame R.

Where D' and E' are the products of inertia of (S') relative to the axes of the reference frame R and G' is the centre of inertia of (S').

(S') is in dynamic equilibrium if G' is on the axis $(O, \vec{z_0})$ and if D'=0 and E'=0.

Let's translate these conditions:

The position of the centre of inertia G' is given by :

$$\overrightarrow{OG'} = \frac{m_1.\overrightarrow{OG} + m_2.\overrightarrow{OM_2} + m_3.\overrightarrow{OM_3}}{m_1 + m_2 + m_3} \qquad (0.42)$$

If G' is on the axis $(O, \vec{z_0})$ this equation is written in projection onto R :

$$\begin{cases} \vec{x} & : \quad m.a + m_2.x_2 + m_3.x_3 = 0 \quad (a) \\ \vec{y} & : \quad m_2.y_2 + m_3.y_3 = 0 \quad (b) \end{cases} \qquad (0.43)$$

The two products of inertia D' and E' have the values :

$$\begin{cases} \vec{x} & : \quad D' = D + m_2.y_2.z_2 + m_3.y_3.z_3 \\ \vec{y} & : \quad E' = D + m_2.x_2.z_2 + m_3.x_3.z_3 \end{cases} \qquad (0.44)$$

If D' and E' are zero, we obtain the following two relationships:

$$\begin{cases} \vec{x} & : \quad D + m_2.y_2.z_2 + m_3.y_3.z_3 = 0 \quad (a) \\ \vec{y} & : \quad D + m_2.x_2.z_2 + m_3.x_3.z_3 = 0 \quad (b) \end{cases} \qquad (0.45)$$

We have four equations (3.43) and (3.45) to determine the eight unknowns *xi, yi, zi et mi* (i=2 and 3). The problem has an infinite number of solutions, so it is necessary to set four conditions in order to solve it. In the case of the dynamic balancing of a vehicle wheel, these conditions are generally linked to the fixing of the masses (balancing weights) on the edge of the rim, on each side of the wheel.

Example: a wheel

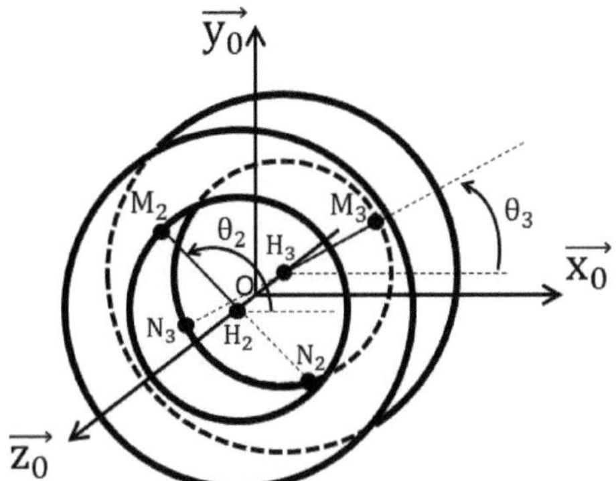

Figure 0.4 Modelling of a vehicle wheel

Let Hi be the orthogonal projection of the point Mi onto the axis $(O, \overrightarrow{z_0})$ and let :

$$\begin{cases} \theta_i = \left(\vec{x}, \overrightarrow{H_iM_i}\right) \\ r_i = \left\|\overrightarrow{H_iM_i}\right\| \end{cases} \tag{0.46}$$

Replace the Cartesian coordinates x_i, y_i, z_i of point Mi with the cylindrical coordinates r_i, θ_i, z_i. The four conditions imposed are the values of the parameters z_2, z_3, r_2, r_3 ($r_2 = r_3$) and the four unknowns are $m_1, m_2, \theta_2, \theta_3$. The unknowns are determined using the four equations above.

By eliminating $m_3.x_3$ between equations (3.43-a) and (3.45-b) we obtain the relationship :

$$m_2.x_2.(z_3 - z_2) = E - m_1.a.z_3 \tag{0.47}$$

By eliminating $m_3.y_3$ between equations (3.43-b) and (3.45-a) we obtain the relationship :

$$m_2.y_2.(z_3 - z_2) = D \tag{0.48}$$

Now $x_2 = r_2.\cos\theta_2$ and $y_2 = r_2.\sin\theta_2$

By replacing x_2 and y_2 by their expressions :

$$m_2.r_2.\cos\theta_2 .(z_3 - z_2) = E - m_1.a.z_3 \tag{0.49}$$

$$m_2.r_2.\sin\theta_2.(z_3 - z_2) = D \tag{0.50}$$

The product of the two equations gives :

$$(E - m_1.a.z_3).\sin\theta_2 = D.\cos\theta_2$$

If $D \neq 0$

$$cotg\theta_2 = \frac{E - m_1.a.z_3}{D}$$

Hence θ_2 modulo π. The exact value of θ_2 is fixed with one of the two equations.

Knowing θ_2 we immediately find m_2. Then by determining θ_3 and m_3 by equations (3.43) and (3.45)

5. Equations of motion (derived from PFD)

The position of a material system in $\mathbf{R_G}$ depends on n parameters q_i (t). The six equations derived from the PFD can be broken down into three scalar equations of the form $f(q_i(t), \dot{q}_i(t), \ddot{q}_i(t)) = 0$.

To simplify the analytical expression of the scalar equations, it is important to choose the reference frame carefully, as well as the point at which to calculate the resultant and the dynamic moment.

These scalar equations are generally second-order differential equations. They contain :

- The system's geometric data
- Inertia characteristics
- The components of mechanical actions (known or unknown)

Equation for a rod

Consider a homogeneous rod of length l, negligible thickness and mass m pivotally connected to the frame along axis $(O, \vec{z})$ with the frame and contained in the plane $(O, \vec{x}, \vec{y})$ (Figure 3.5). Determine the equation of motion of the centre of mass G.

$$\{D_{T/R}\}_0 = \left\{ \begin{matrix} -\frac{m.l}{2}\dot{\alpha}^2\cos\alpha - \frac{m.l}{2}\ddot{\alpha}\sin\alpha & 0 \\ -\frac{m.l}{2}\dot{\alpha}^2\sin\alpha + \frac{m.l}{2}\ddot{\alpha}\cos\alpha & 0 \\ 0 & \frac{m.l^2}{3}\ddot{\alpha} \end{matrix} \right\} \tag{0.51}$$

$$\{T_{pesanteur \to T}\}_0 = \left\{ \begin{matrix} mg & 0 \\ 0 & 0 \\ 0 & -\frac{l.m.g}{2}\sin\alpha \end{matrix} \right\} \tag{0.52}$$

$$\{T_{bati \to T}\}_0 = \left\{ \begin{matrix} X & L \\ Y & M \\ Z & 0 \end{matrix} \right\} \tag{0.53}$$

Application of the PDF gives six scalar equations:

$$\{D_{T/R}\}_A = \{T_{\bar{T} \to T}\}_A \tag{0.54}$$

We find that :

$$\frac{m.l^2}{3}\ddot{\alpha} = -\frac{l.m.g}{2}\sin\alpha \qquad (0.55)$$

Hence the equation of motion is :

$$\ddot{\alpha} + \frac{3.g}{2.l}\sin\alpha = 0 \qquad (0.56)$$

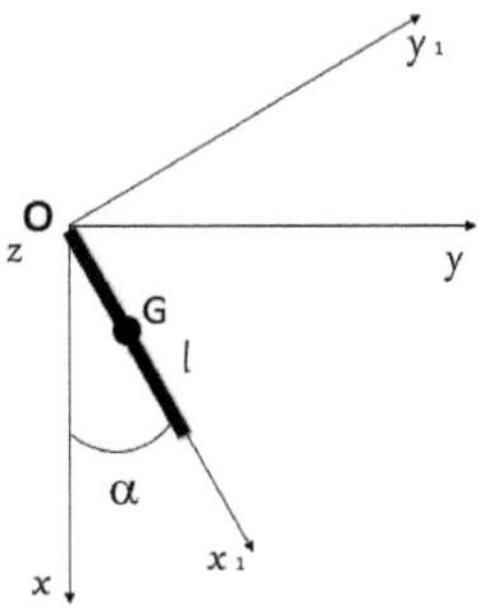

Figure 0.5 Simple pendulum

CHAPTER 4 : Energy

4. Power

4.1. Power developed by mechanical action on a material system

Given two material systems (Σ_1) and (Σ_2), in motion relative to a reference frame R let us assume that (Σ_1) exerts on (Σ_2) a mass density of force $\vec{f}(M)$.

Then the power developed at date t by the mechanical action of (Σ_1) on (Σ_2) in the movement of (Σ_2) relative to R is :

$$P_{(\Sigma1\to\Sigma2/R)} = \int_{M\in\Sigma2} \vec{f}(M).\vec{V}_{M/R}\, dm \tag{0.1}$$

Note:

If the mechanical action of (Σ 1) on (Σ 2) is a force $\vec{F}_M(\Sigma_1 \to \Sigma_2)$ then :

$$P_{(\Sigma1\to\Sigma2/R)} = \vec{F}_{M\,(\Sigma1\to\Sigma2)}.\vec{V}_{M\in\Sigma2/R} \tag{0.2}$$

The unit of power is the Watt (Nms)$^{-1}$

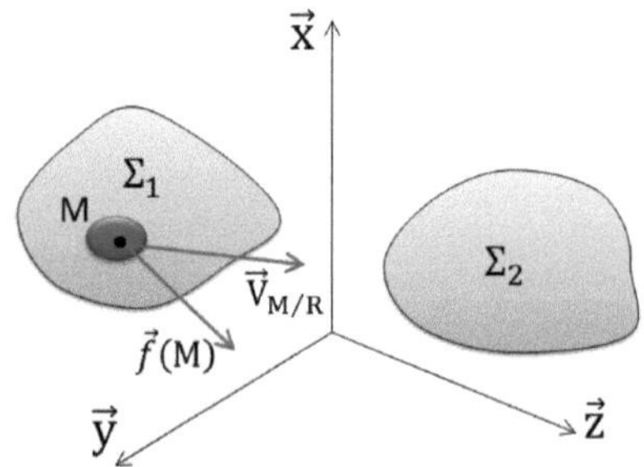

Figure 0.1 Modelling of the mechanical action of (Σ 1) on (Σ 2)

4.2. Power developed by a mechanical action on a solid

Consider that (Σ_2) is a solid (S). We can define $\{V_{S/R}\}$ which is the kinematic torsor defined at a point A on (S).

$$\{V_{S/R}\} = \left\{ \begin{array}{c} \vec{\Omega}_{S/R} \\ \vec{V}_{A\in S/R} \end{array} \right\}_A \tag{0.3}$$

Then the velocity vector of any point M on (S) is expressed as a function of $\vec{V}_{A\in S/R}$ by the relation :

$$\vec{V}_{M\in S/R} = \vec{V}_{A\in S/R} + \vec{\Omega}_{S/R} \wedge \overrightarrow{AM} \tag{0.4}$$

Let's replace this vector in the expression of the power developed by the mechanical action of (Σ_2) on (S), in the movement of (S) relative to R :

$$P_{(\Sigma_1\to\Sigma_2/R)} = \int_{M\in\Sigma_2} \vec{f}(M).\vec{V}_{M/R}\, dm \tag{0.5}$$

$$P_{(\Sigma_1 \to \Sigma_2/R)} = \int\limits_{M \in \Sigma_2} \vec{f}(M).\vec{V}_{A \in S/R}\, dm + \int\limits_{M \in \Sigma_2} \vec{f}(M).\left(\overrightarrow{MA} \wedge \vec{\Omega}_{S/R}\right) dm \qquad (0.6)$$

$$P_{(\Sigma_1 \to \Sigma_2/R)} = \vec{V}_{A \in S/R}.\int\limits_{M \in \Sigma_2} \vec{f}(M)dm + \vec{\Omega}_{S/R}.\int\limits_{M \in \Sigma_2} \overrightarrow{AM} \wedge \vec{f}(M).\,dm \qquad (0.7)$$

If $\{T_{\Sigma_1 ->S}\}$ is the torsor of the mechanical actions of Σ_1 on Σ_2

$$\{T_{\Sigma_1 ->S}\} = \left\{ \frac{\vec{R}}{\overrightarrow{M_A}} \right\} \qquad (0.8)$$

Hence the power developed by the mechanical action of $\Sigma_1 -> S$ in the movement of (S) relative to R is given by :

$$P_{(\Sigma_1 \to S\,/R)} = \{T_{\Sigma_1 ->S}\} \otimes \{V_{S/R}\} \qquad (0.9)$$

Application:

Consider a four-wheeled vehicle travelling in a straight line on a road modelled by the plane $(O\,\vec{y}, \vec{z})$. Let R be a reference frame related to the road (S_0). Suppose that the rear driving wheel (S), with centre A, radius a and negligible thickness, rolls without slipping at point I on the axis $(O\,\vec{y})$. (S) to an assumed perfect pivot link of axis $(A, \vec{z})$ with the chassis (S_1). The drive shaft (S_2) drives (S) in rotation, transmitting to it only a torque of moment $C\vec{z}$.

Consider the following kinematic torsors, expressed in the base R :

$$\{\upsilon(S_1/S_0)\} = \left\{ \begin{matrix} 0 & 0 \\ 0 & v \\ 0 & 0 \end{matrix} \right\}_A \text{ and } \{\upsilon(S/S_1)\} = \left\{ \begin{matrix} 0 & 0 \\ 0 & 0 \\ \omega & 0 \end{matrix} \right\}_A$$

As well as the mechanical action torsors :

$$\{T_{S_0 ->S}\}_I = \left\{ \begin{matrix} N & 0 \\ T & 0 \\ 0 & 0 \end{matrix} \right\}_I \; ; \{T_{S->S_1}\}_A = \left\{ \begin{matrix} X & L \\ Y & M \\ Z & 0 \end{matrix} \right\}_A \; ; \{T_{S_2->S}\}_A = \left\{ \begin{matrix} 0 & 0 \\ 0 & 0 \\ 0 & C \end{matrix} \right\}_A$$

Figure 0.2 Power developed by the action of a wheel on the road

Question :

Determine the power developed by the mechanical action of

1- (S2) on (S) in the movement of (S) relative to (S1)

2- (S0) on (S) in the movement of (S) relative to (S0)

3- (S0) on (S) in the movement of (S) relative to (S1)

4- (S) on (S1) in the movement of (S1) relative to (S0)

5- (S) on (S1) in the movement of (S1) relative to (S)

<u>Response:</u>

These different powers can be calculated as the products of the following torsors, taking care to express them at the same point:

1- $P_{(S_2 \to S\,/S_1)} = \{T_{S_2 \to S}\} \otimes \{V_{S/S_1}\} \Rightarrow P_{(S_2 \to S\,/S_1)} = \begin{Bmatrix} 0 & 0 \\ 0 & 0 \\ 0 & C \end{Bmatrix}_A \otimes \begin{Bmatrix} 0 & 0 \\ 0 & 0 \\ \omega & 0 \end{Bmatrix}_A = C.\,\omega$

2- $P_{(S_0 \to S\,/S_0)} = \{T_{S_0 \to S}\} \otimes \{V_{S/S_0}\} = \vec{R}(S_0 \to S).\vec{V}_{I \in S/S_0}$

As (S) rolls without sliding on (S0), $\vec{V}_{I \in S/S_0}$ is zero, and : $P_{(S_0 \to S\,/S_0)} = 0$

3- $P_{(S_0 \to S\,/S_1)} = \{T_{S_0 \to S}\} \otimes \{V_{S/S_1}\} \Rightarrow P_{(S_0 \to S\,/S_1)} = \begin{Bmatrix} N & 0 \\ T & 0 \\ 0 & 0 \end{Bmatrix}_I \otimes \begin{Bmatrix} 0 & -a.\,\omega \\ \omega & 0 \end{Bmatrix}_I =$

$-a.\,\omega.\,T$

4- $P_{(S \to S_1\,/S_0)} = \{T_{S \to S_1}\} \otimes \{V_{S_1/S_0}\} \Rightarrow P_{(S \to S_1\,/S_0)} = \begin{Bmatrix} X & L \\ Y & M \\ Z & 0 \end{Bmatrix}_A \otimes \begin{Bmatrix} 0 & 0 \\ 0 & v \\ 0 & 0 \end{Bmatrix}_A = M.\,v$

5- $P_{(S \to S_1\,/S)} = \{T_{S \to S_1}\} \otimes \{V_{S_1/S}\} \Rightarrow P_{(S \to S_1\,/S)} = \begin{Bmatrix} X & L \\ Y & M \\ Z & 0 \end{Bmatrix}_A \otimes \begin{Bmatrix} 0 & 0 \\ 0 & 0 \\ -\omega & 0 \end{Bmatrix}_A = 0$

4.3. Power of mutual actions between two solids

Let us consider two solids, S_1 and S_2 , moving relative to each other and both moving relative to a reference frame R. In this case, S_1 and S_2 are subject to mechanical actions defined by a linkage torsor. Taking into account the definition of the power of the mechanical actions, we can write for S_2 :

$$P_{(S_2 \to S_1\,/R)} = \{T_{S_2 \to S_1}\} \otimes \{V_{S_1/R}\} \tag{0.10}$$

If we consider the composition of kinematic torsors, we can write for S :$_1$

$$P_{(S_1 \to S_2\,/R)} = \{T_{S_1 \to S_2}\} \otimes \{V_{S_2/R}\} = \{T_{S_1 \to S_2}\} \otimes (\{V_{S_2/S_1}\} + \{V_{S_1/R}\}) \tag{0.11}$$

If we add the two previous relations (4.10 and 4.11) term by term, we obtain :

$$P_{(S_1 \to S_2\,/R)} + P_{(S_2 \to S_1\,/R)} = \{T_{S_1 \to S_2}\} \otimes \{V_{S_2/S_1}\} \tag{0.12}$$

$$P_{(S_1 \to S_2\,/R)} + P_{(S_2 \to S_1\,/R)} = P_{(S_1 \leftrightarrow S_2\,/R)} \tag{0.13}$$

$P_{(S_1 \leftrightarrow S_2\,/R)}$ is the power developed by the mutual actions between (S_1) and (S_2), and is independent of the reference frame.

<u>Demo :</u>

$$P_{(S_1 \to S_2 /R)} - P_{(S_1 \to S_2 /R_1)} = \{T_{S_1 \to S_2}\} \otimes \{V_{S_2/R}\} - \{T_{S_1 \to S_2}\} \otimes \{V_{S_2/R_1}\} \qquad (0.14)$$

$$P_{(S_1 \to S_2 /R)} - P_{(S_1 \to S_2 /R_1)} = \{T_{S_1 \to S_2}\} \otimes \{V_{R_1/R}\} \qquad (0.15)$$

$$P_{(S_2 \to S_1 /R)} - P_{(S_2 \to S_1 /R_1)} = \{T_{S_2 \to S_1}\} \otimes \{V_{S_1/R}\} - \{T_{S_2 \to S_1}\} \otimes \{V_{S_1/R_1}\} \qquad (0.16)$$

$$P_{(S_2 \to S_1 /R)} - P_{(S_2 \to S_1 /R_1)} = \{T_{S_2 \to S_1}\} \otimes \{V_{R_1/R}\} \qquad (0.17)$$

The sum of the two relations (4.15) and (4.17) gives :

$$P_{(S_1 \Leftrightarrow S_2 /R)} - P_{(S_2 \Leftrightarrow S_1 /R_1)} = \left(\{T_{S_1 \to S_2}\} + \{T_{S_2 \to S_1}\}\right) \otimes \{V_{R_1/R}\} \qquad (0.18)$$

According to the mutual action theorem, the second member of (4.18) is zero. Therefore :

$$P_{(S_1 \Leftrightarrow S_2 /R)} = P_{(S_2 \Leftrightarrow S_1 /R_1)} \qquad (0.19)$$

This proves that the power developed by the mutual actions between (S_1) and (S_2) is independent of the choice of reference frame used to calculate it. Consequently this power will simply be noted as :

$$\mathbf{P(S1 \Longleftrightarrow S2)} \qquad (0.20)$$

5. Work

5.1. Definition

The work of the mechanical action exerted by the material assembly (S_1) on the material assembly (S_2), during the displacement of (S_2) relative to the reference frame **R**, between the instants t_1 and t_2 , is :

$$W_{t_1}^{t_2}(S_1 \to S_2 /R) = \int_{t_1}^{t_2} P(S_1 \to S_2 /R)\, dt$$

As a result, the elementary work between the dates t and $t + dt$ is :

$$dW(S_1 \to S_2 /R) = P(S_1 \to S_2 /R).\, dt$$

5.2. Units

- The unit of work is the joule.
- The unit of power is the watt.

6. Kinetic energy :

6.1. Definition

The kinetic energy of a system (S) in its motion relative to a fixed reference frame R is defined by the scalar :

$$E_C = \frac{1}{2} \cdot \int \left[\vec{V}_{M \in S / R_g} \right]^2 \cdot dm \tag{0.21}$$

If (S) is a material system with centre of gravity G and mass m in motion relative to a reference frame Rg, it can be written at any point A.

$$\left\{ D_{S/Rg} \right\}_A = \left\{ T_{\bar{S} \to S} \right\}_A \tag{0.22}$$

Let's multiply the two members of this relationship by the kinematic torsor of (S) with respect to Rg:

$$\left\{ D_{S/Rg} \right\}_A \cdot \left\{ v_{S/Rg} \right\}_A = \left\{ T_{\bar{S} \to S} \right\}_A \cdot \left\{ v_{S/Rg} \right\}_A \tag{0.23}$$

With $\left\{ T_{\bar{S} \to S} \right\}_A \cdot \left\{ v_{S/Rg} \right\}_A$ is the power of mechanical actions external to (S) exerted on (S) in its movement relative to Rg

The result is :

$$\left\{ \begin{array}{c} \int \vec{\Gamma}_{M/R_g} \, dm \\ \int \overrightarrow{AM} \wedge \vec{\Gamma}_{M/R_g} dm \end{array} \right\}_A \cdot \left\{ \begin{array}{c} \vec{\Omega}_{S/R_g} \\ \vec{V}_{A \in S/R_g} \end{array} \right\}_A = \left\{ \begin{array}{c} \vec{R}_{(\bar{S} \to S)} \\ \vec{M}_{(\bar{S} \to S)} \end{array} \right\}_A \cdot \left\{ \begin{array}{c} \vec{\Omega}_{S/R_g} \\ \vec{V}_{A \in S/R_g} \end{array} \right\}_A = P(\bar{S} \to S / R_g) \tag{0.24}$$

$$P(\bar{S} \to S / R_g) = \vec{V}_{A \in S/R_g} \cdot \int \vec{\Gamma}_{M/R_g} \, dm + \vec{\Omega}_{S/R_g} \cdot \int \overrightarrow{AM} \wedge \vec{\Gamma}_{M/R_g} dm \tag{0.25}$$

The velocity vector of a point M on (S) is expressed as a function of $\vec{V}_{A \in S/R}$ by the relation :

$$\vec{V}_{M \in S/R_g} = \vec{V}_{A \in S/R_g} + \vec{\Omega}_{S/R_g} \wedge \overrightarrow{AM} \tag{0.26}$$

So

$$\vec{V}_{A \in S/R_g} \cdot \int \vec{\Gamma}_{M/R} \, dm = \int \vec{\Gamma}_{M/R_g} \cdot \vec{V}_{M \in S/R_g} \, dm - \int \vec{\Gamma}_{M/R_g} \cdot (\vec{\Omega}_{S/R_g} \wedge \overrightarrow{AM}) \, dm \tag{0.27}$$

Since :

$$P(\bar{S} \to S / R_g) = \vec{\Gamma}_{M/R} \cdot \vec{V}_{M \in S/R_g} = \frac{d}{dt} \left[\vec{V}_{M \in S/R_g} \right] \cdot \vec{V}_{M \in S/R_g} = \frac{1}{2} \cdot \frac{d}{dt} \left[\vec{V}_{M \in S/R_g} \right]^2$$

$$\tag{0.28}$$

We can therefore write :

$$\int \vec{\Gamma}_{M/R} \cdot \vec{V}_{M \in S/R_g} \cdot dm = \frac{d}{dt} \left(\frac{1}{2} \cdot \int \left[\vec{V}_{M \in S/R_g} \right]^2 \cdot dm \right) \tag{0.29}$$

This gives the kinetic energy of (S) as it moves relative to Rg :

$$E_C = \frac{1}{2} \cdot \int \left[\vec{V}_{M \in S/R_g} \right]^2 \cdot dm \tag{0.30}$$

As a result, the kinetic energy theorem is written for a solid (S):

$$P(\bar{S} \to S / R_g) = \frac{d}{dt} E_C(S / R_g) \tag{0.31}$$

Kinetic energy is a SCALAIRE, unit (W.s)

6.2. Kinetic energy of translation and rotation

Let (R_0) be a fixed reference frame and (R_S) a moving reference frame linked to the solid (S). Consider the point G which represents the centre of mass of (S) and M any point on (S). In this case, we can write :

$$\vec{V}_{M\in S/R} = \vec{V}_{G\in S/R} + \overrightarrow{MG} \wedge \vec{\Omega}_{S/R} \tag{0.32}$$

Hence

$$\left(\vec{V}_{M\in S/R}\right)^2 = \left(\vec{V}_{G\in S/R}\right)^2 + \left(\overrightarrow{MG} \wedge \vec{\Omega}_{S/R}\right)^2 + 2.\vec{V}_{G\in S/R}.\overrightarrow{MG} \wedge \vec{\Omega}_{S/R} \tag{0.33}$$

From the mixed product property: $((\vec{A} \wedge \vec{B}).\vec{C} = \vec{A}.(\vec{B} \wedge \vec{C}) = \vec{B}.(\vec{C} \wedge \vec{A}))$ we obtain :

$$\left(\vec{V}_{M\in S/R}\right)^2 = \left(\vec{V}_{G\in S/R}\right)^2 + 2.\vec{V}_{G\in S/R}.\overrightarrow{PG} \wedge \vec{\Omega}_{S/R} + \vec{\Omega}_{S/R}.\left(\left(\overrightarrow{MG} \wedge \vec{\Omega}_{S/R}\right) \wedge \overrightarrow{MG}\right)$$
$$\tag{0.34}$$

Using the properties of vector associativity and the invariance of $\vec{\Omega}_{S/R}$ and $\vec{V}_{G\in S/R}$ with respect to the integration element dm, we can write

$$E\left(\frac{S}{R}\right) = \frac{1}{2}\int\left(\vec{V}_{M/R}\right)^2 dm = \frac{1}{2}\left(\vec{V}_{G\in S/R}\right)^2 \int dm + 2.\vec{V}_{G\in S/R}.\vec{\Omega}_{S/R} \wedge \int \overrightarrow{GM}.dm +$$
$$\vec{\Omega}_{S/R}.\int\left(\overrightarrow{MG} \wedge \vec{\Omega}_{S/R}\right) \wedge \overrightarrow{MG}\, dm \tag{0.35}$$

We've got more:

$$\int dm = M \text{ and } \int \overrightarrow{GM}.dm = \vec{0}$$

$$\int\left(\overrightarrow{MG} \wedge \vec{\Omega}_{S/R}\right) \wedge \overrightarrow{MG}\, dm = \int \overrightarrow{GM} \wedge \left(\vec{\Omega}_{S/R} \wedge \overrightarrow{GM}\right) dm = [I_G(s)].\vec{\Omega}_{S/R} \tag{0.36}$$

Hence, in the final analysis :

$$E(S/R) = \frac{1}{2}M.\vec{V}_{G\in S/R}^{\;2} + \frac{1}{2}[I_G(s)].\vec{\Omega}_{S/R}^{\;2} \tag{0.37}$$

- $\frac{1}{2}M.\vec{V}_{G\in S/R}^{\;2}$ corresponds to the translational kinetic energy of the solid that would be obtained if G had rectilinear translational motion.

- $\frac{1}{2}[I_G(s)].\vec{\Omega}_{S/R}^{\;2}$ corresponds to the kinetic energy of rotation that would be obtained if G were fixed.

4.4. Application

Kinetic energy of a rod

Consider a homogeneous rod of length l, negligible thickness and mass m in pivotal connection of axis $(O, \vec{z})$ with the frame and contained in the plane $(O, \vec{x}, \vec{y})$ (Figure 3.5). Determine the kinetic energy of the rod (T) in its motion relative to R.

$$E_C = \frac{1}{2}.\int\left[\vec{V}_{M\in S/R_g}\right]^2.dm \tag{0.38}$$

$$\vec{V}_{M \in S /R_g} = \left[\frac{d}{dt}\overrightarrow{OM}\right]_R = x\dot{\alpha}\,\overrightarrow{y_1} \tag{0.39}$$

$$E_C = \frac{1}{2}\cdot\int(x\dot{\alpha}\,\overrightarrow{y_1})^2\cdot dm = \frac{m.\dot{\alpha}^2}{2.l}\cdot\int_0^l x^2\cdot dx = \frac{m.\dot{\alpha}^2.l^2}{6} \tag{0.40}$$

Or :

$$Ec = \frac{1}{2}m.\vec{V}_{G\in S/R}^{\,2} + \frac{1}{2}[I_G(s)].\vec{\Omega}_{S/R}^{\,2} \tag{0.41}$$

$$Ec = \frac{1}{2}m.\left(\frac{l.\dot{\alpha}}{2}\right)^2 + \frac{1}{2}\left[\frac{m.l^2}{12}\right].\dot{\alpha}^2 = \frac{m.l^2.\dot{\alpha}^2}{8} + \frac{m.l^2.\dot{\alpha}^2}{24} = \frac{3.m.l^2.\dot{\alpha}^2}{24} + \frac{m.l^2.\dot{\alpha}^2}{24} = \frac{m.l^2.\dot{\alpha}^2}{6} \tag{0.42}$$

7. Potential energy

7.1. Definition

The material assembly ($\Sigma2$) has a potential energy $E_{P_{(\Sigma1\to\Sigma2/R)}}$ associated with the mechanical

action of ($\Sigma1$) on ($\Sigma2$), in the motion of ($\Sigma2$) relative to R. It is expressed by :

$$P_{(\Sigma1\to\Sigma2/R)} = -\frac{d}{dt}E_{P_{(\Sigma1\to\Sigma2/R)}} \tag{0.43}$$

7.2. Potential energy of two sets of materials, associated with mutual action

The two material sets ($\Sigma1$) and ($\Sigma2$) have potential energy, associated with mutual action, if

there exists a scalar function $E_{P_{(\Sigma1\leftrightarrow\Sigma2)}}$ such that :

$$P_{(\Sigma1\leftrightarrow\Sigma2)} = -\frac{d}{dt}E_{P_{(\Sigma1\leftrightarrow\Sigma2)}} \tag{0.44}$$

7.3.1. Example: *potential energy of two solids with a tension-compression spring of zero*

mass between them.

Consider two solids (S_1) and (S_2) have a sliding pivot connection of axis $(O,\vec{\imath})$. A tension-compression spring (r) of mass assumed to be zero, stiffness K, no-load length "$\boldsymbol{\ell}$" placed along the axis $(O,\vec{\imath})$ between (S_1) and (S_2). The mechanical action of the spring (r) on (S_2) is modelled by the torsor $\{T_{r->S_2}\}$.

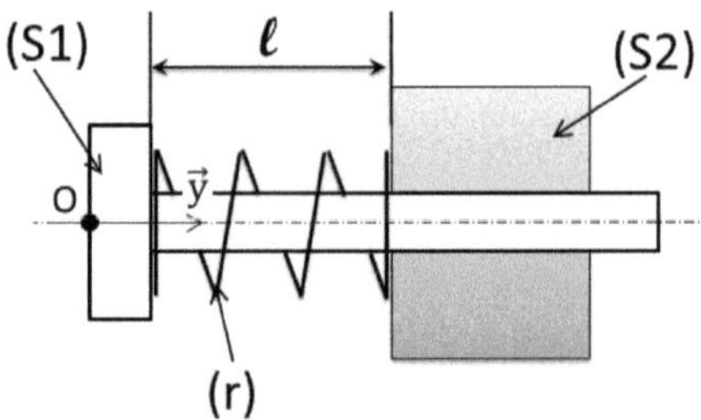

Figure 0.3 Two solids with a spring between them

Let's calculate the power produced by the mechanical action of the spring (r) on the two solids (S1) and (S2) as they move relative to any reference frame R.

$$P(r \to S_1, S_2/R) = P(r \to S_1/R) + P(r \to S_2/R)$$

Or

$$P(r \rightarrow S_1, S_2/R) = \{T_{r->S_1}\}.\{V_{S_1/R}\} + \{T_{r->S_2}\}.\{V_{S_2/R}\}$$

The spring has zero mass:

$$\{T_{r->S_1}\} = -\{T_{r->S_2}\}$$

Hence

$$P(r \rightarrow S_1, S_2/R) = \{T_{r->S_2}\}.\left(\{V_{S_2/R}\} - \{V_{S_1/R}\}\right)$$

Or

$$P(r \rightarrow S_1, S_2/R) = \{T_{r->S_2}\}.\{V_{S_2/S_1}\}$$

This power is independent of the reference frame R. We will simply write it down as :

$$P(r \rightarrow S_1, S_2) = -K(l - l_0).\frac{dl}{dt}$$

There is therefore a potential energy of (S_1) and (S_2) associated with their mutual action via the spring (r):

$$E_p(r \rightarrow S_1, S_2) = \frac{K}{2}(l - l_0)^2$$

7.3.2. Example 2: _Potential energy of a rod_

Consider a homogeneous rod of length l, negligible thickness and mass m in pivotal connection of axis $(O, \vec{z})$ with the frame and contained in the plane $(O, \vec{x}, \vec{y})$ (Figure 3.5). Determine the potential energy of the rod (T) in its motion relative to R.

$$P_{(\overline{T}\rightarrow T/R)} = P_{(bati\rightarrow T/R)} + P_{(Pesanteur\rightarrow T/R)} \tag{0.45}$$

The connection between (T) and the frame is perfect, so

$$P_{(bati\leftrightarrow T)} = 0 \tag{0.46}$$

Since R is connected to the frame

$$P_{(bati\rightarrow T/R)} = P_{(bati\leftrightarrow T)} = 0 \tag{0.47}$$

$$P_{(\overline{T}\rightarrow T/R)} = P_{(Pesanteur\rightarrow T/R)} = \{T_{Pesanteur\rightarrow T}\} \otimes \{V_{T/R}\} \tag{0.48}$$

$$P_{(Pesanteur\rightarrow T/R)} = \left\{\begin{matrix} mg & 0 \\ 0 & 0 \\ 0 & -\frac{l.m.g}{2}\sin\alpha \end{matrix}\right\} \otimes \left\{\begin{matrix} 0 & 0 \\ 0 & 0 \\ \dot{\alpha} & 0 \end{matrix}\right\} = -\frac{l.m.g}{2}.\dot{\alpha}.\sin\alpha \tag{0.49}$$

Gold

$$P_{(Pesanteur\rightarrow T/R)} = -\frac{d}{dt}E_{p(Pesanteur\leftrightarrow T)} = -\frac{l.m.g}{2}\frac{d}{dt}(-\cos\alpha) \tag{0.50}$$

Given the previous relationship

$$E_{p(Pesanteur\leftrightarrow T)} = -\frac{l.m.g}{2}.\cos\alpha \tag{0.51}$$

According to relations (4.41) and (4.49)

$$P_{(\overline{T}\to T/R)} = \frac{d}{dt}E_C(T/R_g) = -\frac{d}{dt}E_{P_{(Pesanteur\leftrightarrow T)}} \tag{0.52}$$

$$\frac{d}{dt}\left(\frac{m.\dot{\alpha}^2.l^2}{6}\right) = -\frac{d}{dt}\left(-\frac{l.m.g}{2}.\cos\alpha\right) \tag{0.53}$$

$$\frac{d}{dt}\left(\frac{m.\dot{\alpha}^2.l^2}{6} - \frac{l.m.g}{2}.\cos\alpha\right) = 0 \tag{0.54}$$

$$\frac{m.l^2}{3}\dot{\alpha}.\left(\ddot{\alpha} + \frac{3.g}{2.l}.\sin\alpha\right) = 0 \tag{0.55}$$

CHAPTER 5 : Corrected problems

5. The circular saw

5.1.Statement

We propose to study the circular saw used to cut tubes (Figure 1). It consists of :

- A frame (S_0) with a supposedly Galilean reference frame R_0 $(O, \vec{x_0}, \vec{y_0}, \vec{z_0})$.

- An arm (S_1), in perfect pivot connection of axis $(O, \vec{z_0})$ with the frame (S_0), to which is attached the reference frame R_1 $(O, \vec{x_1}, \vec{y_1}, \vec{z_0})$ such that $\alpha = (\vec{x_0}, \vec{x_1}) = (\vec{y_0}, \vec{y_1})$ (α is positive).

- An assembly (S_2), in perfect pivot connection of axis $(A, \vec{z_0})$ with the arm (S_1), to which is attached the reference frame R_2 $(A, \vec{x_2}, \vec{y_2}, \vec{z_0})$ such that $\beta = (\vec{x_1}, \vec{x_2}) = (\vec{y_1}, \vec{y_2})$ (α is positive). The assembly (S_2) is composed of :

  - A tree is treated as a homogeneous **stem (T)** of mass m_1, length **L**, negligible radius and <u>centre of mass **A.**</u>

  - A saw assimilated to a homogeneous **disc (D)** of mass m_2, radius **R**, negligible thickness and <u>centre of mass **B**.</u>

- A tube (S_3) clamped to the frame (S_0) during the cutting operation. At the start of this operation, the contact between the assembly (S_2) and the tube (S_3), ensured with friction of coefficient f, is assumed to be punctual of normal $(I, \vec{z_0})$.

All the Ri reference frames used are direct orthonormal. The gravity field is defined by $\vec{g} = - g.\vec{z_0}$. The action of (S_2) on (S_3) at point I is defined, in the reference frame R_0, by the torsor :

$$\{T_{2\rightarrow3}\} = \left\{ \begin{matrix} T & | & 0 \\ N & | & 0 \\ 0 & | & 0 \end{matrix} \right\}_I .$$

The arm (S_1) is driven in rotation by a geared motor M_1 exerting a torque : $\vec{C_1} = C_1.\vec{z_0}$

The assembly (S_2) is driven in rotation by a motor M_2 exerting a torque: $\vec{C_2} = C_2.\vec{z_0}$

We give : $\vec{OA} = r.\vec{x_1}$; $\vec{AB} = \frac{L}{2}.\vec{z_0}$; $\vec{IB} = R.\vec{y_0}$

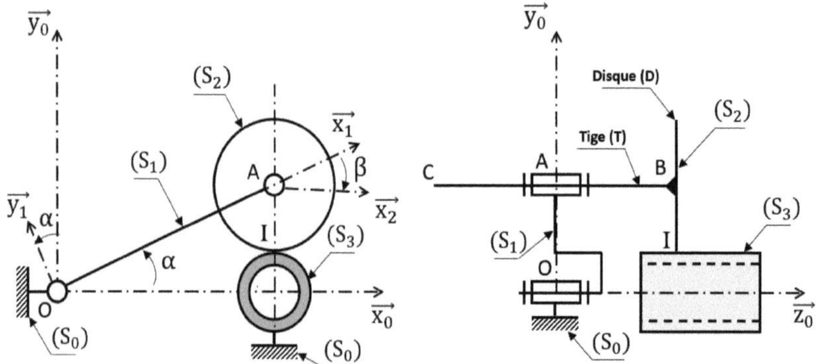

Figure 1: Minimum kinematic diagram of the chainsaw

1) In the reference frame R_2 $(A, \vec{x_2}, \vec{y_2}, \vec{z_0})$ the position of the centre of inertia G of the assembly (S_2):

$$M.\overrightarrow{AG} = \sum_{i=1}^{n} \overrightarrow{AG_i}.m_i \quad \text{and} \quad M = \sum_{i=1}^{n} m_i$$

2) We propose to add to the set (S_2) a point mass m_3 concentrated at point C. Determine m_3 so that the centre of inertia G coincides with point A : $\overrightarrow{AB} = \overrightarrow{GB} = \frac{L}{2}\vec{z_0}$

3) Show that the inertia matrix of the rod (T) at point A (centre of inertia of the rod) expressed in R_2 is of the form : $[I_A(T)]_{R_2} = \begin{bmatrix} A_T & 0 & 0 \\ 0 & A_T & 0 \\ 0 & 0 & 0 \end{bmatrix}$ then show that $A_T = \int z^2.dm = \frac{m_1.L^2}{12}$

4) Show that the inertia matrix of the disc (D) at point B (centre of inertia of the disc) expressed in R_2 is of the following form: $[I_B(D)]_{R_2} = \begin{bmatrix} A_D & 0 & 0 \\ 0 & A_D & 0 \\ 0 & 0 & C_D \end{bmatrix}$

5) Using Huygens' theorem, show that the inertia matrix of the set $(S)_2 [I_A(S_2)]_{R_2}$ expressed at point (A) is of the following form : $[I_A(S_2)]_{R_2} = \begin{bmatrix} A & 0 & 0 \\ 0 & A & 0 \\ 0 & 0 & C \end{bmatrix}$

6) Express the kinetic torsor of (S_2) with respect to (S_0) at point A.

7) Express the dynamic torsor of (S_2) with respect to (S_0) at point A .

8) In the base R_0 the torsor associated with these actions at point A: $\{T_{\overline{S_2} \to S_2}\}_A$

9) Determine the scalar equations derived from the PFD (Fundamental Principle of Dynamics) applied to the set $((S_2)$ in motion relative to R_0. Derive T and N as a function of C_2.

5.2. Correction

1) In the reference frame R_2 $(A, \overrightarrow{x_2}, \overrightarrow{y_2}, \overrightarrow{z_0})$ the position of the centre of inertia G of the assembly (S_2):

$$\mathbf{M}.\overrightarrow{\mathbf{AG}} = \sum_{i=1}^{n} \overrightarrow{\mathbf{AG_i}}.\mathbf{m_i} \text{ and } \mathbf{M} = \sum_{i=1}^{n} \mathbf{m_i}$$

$$\mathbf{M}.\overrightarrow{\mathbf{AG}} = m_1.\overrightarrow{\mathbf{AA}} + m_2.\overrightarrow{\mathbf{AB}} \text{ with } \mathbf{M} = m_1 + m_2$$

$$\overrightarrow{\mathbf{AA}} = \vec{0} \text{ hence } \overrightarrow{\mathbf{AG}} = \frac{m_2}{M}.\overrightarrow{\mathbf{AB}}$$

we have $\overrightarrow{AB} = \frac{L}{2}.\overrightarrow{z_0}$

so $\overrightarrow{\mathbf{AG}} = \frac{m_2}{M}.\frac{L}{2}.\overrightarrow{z_0} = \frac{m_2}{m_1+m_2}.\frac{L}{2}.\overrightarrow{z_0}$

2) We propose to add a point mass (S_2) a point mass m_3 concentrated at point C. Determine m_3 so that the centre of inertia G coincides with point A :

$$\overrightarrow{AB} = \overrightarrow{GB} = \frac{L}{2}\overrightarrow{z_0}$$

$$\mathbf{M}.\overrightarrow{\mathbf{AG}} = \sum_{i=1}^{n} \overrightarrow{\mathbf{AG_i}}.\mathbf{m_i} \text{ and } A \equiv G$$

$$\mathbf{M}.\overrightarrow{\mathbf{AA}} = m_1.\overrightarrow{\mathbf{AA}} + m_2.\overrightarrow{\mathbf{AB}} + m_3.\overrightarrow{\mathbf{AC}}$$

or $\overrightarrow{AA} = \vec{0}$

hence $m_2.\overrightarrow{AB} + m_3.\overrightarrow{AC} = \vec{0}$

$m_2.\overrightarrow{AB} + m_3.\overrightarrow{AC} = \vec{0}$ we have $\overrightarrow{AB} = \frac{L}{2}.\overrightarrow{z_0}$ and we have $\overrightarrow{AC} = -\frac{L}{2}.\overrightarrow{z_0}$

so $m_2.\frac{L}{2}.\overrightarrow{z_0} - m_3.\frac{L}{2}.\overrightarrow{z_0} = \vec{0}$ we find that $m_2 = m_3$

3) Show that the inertia matrix of the rod (T) at point A (centre of inertia of the rod) expressed in R^2 is of the following form: $[I_A(T)]_{R_2} = \begin{bmatrix} A_T & 0 & 0 \\ 0 & A_T & 0 \\ 0 & 0 & 0 \end{bmatrix}$

$(O, \overrightarrow{Z})$ is an axis of revolution (axis of material symmetry) then the inertia matrix is principal

$$[I_A(T)]_{R_2} = \begin{bmatrix} A_T & 0 & 0 \\ 0 & A_T & 0 \\ 0 & 0 & C_T \end{bmatrix}$$

With $A_T = \int (x^2 + z^2).dm = \int (y^2 + z^2).dm$ (1)

$\qquad C_T = \int (x^2 + y^2).dm$ (2)

the diameter of the rod is negligible, i.e. $x^2 = y^2 = 0$

in this case $A_T = \int (x^2 + z^2).dm = \int z^2.dm$ and $C_T = \int (x^2 + y^2).dm = 0$

$A_T = \int z^2.dm$ if we consider that $dm = \lambda dz = \frac{m_1}{L}dz$

so $A_T = \int z^2.dm = A_T = \int z^2.\frac{m_1}{L}dz = \frac{m_1}{L}\int_{-L/2}^{L/2} z^2.dz = \frac{m_1}{L}.\left[\frac{1}{3}z^3\right]_{-L/2}^{L/2} = \frac{m_1}{L}.\frac{1}{3}.\frac{L^3}{4} = \frac{m_1.L^2}{12}$

4) Show that the inertia matrix of the disc (D) at point B (centre of inertia of the disc) expressed in R_2 is of the following form: $[I_B(D)]_{R_2} = \begin{bmatrix} A_D & 0 & 0 \\ 0 & A_D & 0 \\ 0 & 0 & C_D \end{bmatrix}$

The inertia matrix of (D) relative to point B is given by $[I_B(D)]_{R_2}$ is in the following general form

$$[I_B(D)]_{R_2} = \begin{bmatrix} A & -F & -E \\ -F & B & -D \\ -E & -D & C \end{bmatrix}$$

$(O, \vec{Z})$ is an axis of revolution (axis of material symmetry) then the inertia matrix is principal then F=E=D=0

$$\text{So } [I_B(D)]_{R_2} = \begin{bmatrix} A_D & 0 & 0 \\ 0 & B_D & 0 \\ 0 & 0 & C_D \end{bmatrix}$$

the axes $(o, \vec{x})$ and $(o, \vec{y})$ are equivalent, so $A = B$ in this case we find :

$$[I_B(D)]_{R_2} = \begin{bmatrix} A_D & 0 & 0 \\ 0 & A_D & 0 \\ 0 & 0 & C_D \end{bmatrix}$$

We'll take it from there: $A_T = \frac{m_1.L^2}{12}$, $A_D = \frac{m_2.R^2}{4}$ et $C_D = \frac{m_2.R^2}{2}$

5) Using Huygens' theorem, show that the inertia matrix of the set (S) $_2[I_A(S_2)]_{R_2}$ expressed at point (A) is of the following form : $[I_A(S_2)]_{R_2} = \begin{bmatrix} A & 0 & 0 \\ 0 & A & 0 \\ 0 & 0 & C \end{bmatrix}$

with $A = \frac{m_1.L^2}{12} + \frac{m_2.R^2}{4} + \frac{m_2.L^2}{4}$; $C = \frac{m_2.R^2}{2}$

$$[I_A(S_2)]_{R_2} = [I_A(T)]_{R_2} + [I_A(D)]_{R_2}$$

The expression for the inertia matrix of the disc at point B is given by :

$$[I_B(D)]_{R_2} = \begin{bmatrix} A_D & 0 & 0 \\ 0 & A_D & 0 \\ 0 & 0 & C_D \end{bmatrix}$$

To determine the expression of this matrix at point A we use Huygens' theorem

$$[I_A(D)]_{R_2} = [I_B(D)]_{R_2} + [I(B.A.m_2)]$$

The vector $\overrightarrow{AB} = \frac{L}{2}.\overrightarrow{z_0}$

hence
$$[I(B.A.m_2)] = m_2 \begin{bmatrix} y_B^2 + z_B^2 & -x_B.y_B & -x_B.z_B \\ -x_B.y_B & x_B^2 + z_B^2 & -y_B.z_B \\ -x_B.z_B & -y_B.z_B & x_B^2 + y_B^2 \end{bmatrix} =$$

$$m_2 \begin{bmatrix} 0^2 + \frac{L^2}{4} & 0 & 0 \\ 0 & 0^2 + \frac{L^2}{4} & 0 \\ 0 & 0 & 0^2 + 0^2 \end{bmatrix}$$

$$[I_A(D)]_{R_2} = \begin{bmatrix} A & 0 & 0 \\ 0 & B & 0 \\ 0 & 0 & C \end{bmatrix} = \begin{bmatrix} A_D & 0 & 0 \\ 0 & A_D & 0 \\ 0 & 0 & C_D \end{bmatrix} + m_2 \begin{bmatrix} \frac{L^2}{2} & 0 & 0 \\ 0 & \frac{L^2}{2} & 0 \\ 0 & 0 & 0 \end{bmatrix} = m_2 \begin{bmatrix} \frac{R^2}{4} + \frac{L^2}{4} & 0 & 0 \\ 0 & \frac{R^2}{4} + \frac{L^2}{4} & 0 \\ 0 & 0 & \frac{R^2}{2} \end{bmatrix}$$

the inertia matrix of the set $(S)_2$ $[I_A(S_2)]_{R_2}$ is

$$[I_A(S_2)]_{R_2} = [I_A(T)]_{R_2} + [I_A(D)]_{R_2}$$

$$[I_A(S_2)]_{R_2} = \begin{bmatrix} m_1\frac{L^2}{12} & 0 & 0 \\ 0 & m_1\frac{L^2}{12} & 0 \\ 0 & 0 & 0 \end{bmatrix} + \begin{bmatrix} \frac{m_2.R^2}{4} + m_2\frac{L^2}{4} & 0 & 0 \\ 0 & \frac{m_2.R^2}{4} + m_2\frac{L^2}{4} & 0 \\ 0 & 0 & \frac{m_2.R^2}{2} \end{bmatrix}$$

$$[I_A(S_2)]_{R_2} = \begin{bmatrix} m_1\frac{L^2}{12} + \frac{m_2.R^2}{4} + m_2\frac{L^2}{4} & 0 & 0 \\ 0 & m_1\frac{L^2}{12} + \frac{m_2.R^2}{4} + m_2\frac{L^2}{4} & 0 \\ 0 & 0 & \frac{m_2.R^2}{2} \end{bmatrix}$$

Let $\{C_{S2\to S0}\}_A$ is the kinematic torsor representing the movement of (S_2) relative to (S_0) at point **A,** <u>the centre of mass of the assembly (S_2)</u>;

$$\{C_{S2\to S0}\}_A = \left\{ \begin{array}{l} \vec{\Omega}(S_2/S_0) = (\dot{\alpha} + \dot{\beta})\vec{z_0} \\ \vec{V}(A \in S2/R_0) = r.\dot{\alpha}\,\vec{y_1} \end{array} \right\}_A$$

6) Express the kinetic torsor of (S_2) with respect to (S_0) at point A.

$$\{C(S_2/R_0)\}_A = \left\{ \begin{array}{l} \vec{Q} = M\,\vec{V}_{B\in S2/R0} \\ \vec{\sigma_A}(S_2/R_0) = [I_A(S_2)].\vec{\Omega}(S_2/S_0) \end{array} \right\}$$

$$\vec{Q} = M\,\vec{V}_{B\in S2/R0} = (m_1 + m_2)r.\dot{\alpha}\,\vec{y_1} = (m_1 + m_2)r.\dot{\alpha}\,(c\alpha\,\vec{x_0} + s\alpha\,\vec{y_0})$$

$$\vec{\sigma_A}(S_2/R_0) = [I_A(S_2)].\vec{\Omega}\left(\frac{S_2}{S_0}\right) =$$

$$\begin{bmatrix} m_1\frac{L^2}{12} + \frac{m_2.R^2}{4} + m_2\frac{L^2}{4} & 0 & 0 \\ 0 & m_1\frac{L^2}{12} + \frac{m_2.R^2}{4} + m_2\frac{L^2}{4} & 0 \\ 0 & 0 & \frac{m_2.R^2}{2} \end{bmatrix} \cdot \begin{pmatrix} 0 \\ 0 \\ \dot{\alpha} + \dot{\beta} \end{pmatrix}$$

$$\vec{\sigma_A}(S_2/R_0) = \frac{m_2.R^2}{2}(\dot{\alpha} + \dot{\beta})\vec{z_0}$$

$$\{C(S_2/R_0)\}_A = \left\{ \begin{array}{l} \vec{Q} = M\,\vec{V}_{B\in S2/R0} \\ \vec{\sigma_A}(S_2/R_0) = [I_A(S_2)].\vec{\Omega}(S_2/S_0) \end{array} \right\} = \left\{ \begin{array}{l|l} (m_1 + m_2)r.\dot{\alpha}.c\alpha & 0 \\ (m_1 + m_2)r.\dot{\alpha}.s\alpha & 0 \\ 0 & \frac{m_2.R^2}{2}(\dot{\alpha} + \dot{\beta}) \end{array} \right\}$$

7) Express the dynamic torsor of (S_2) with respect to (S_0) at point A.

$$\{D_{\overrightarrow{\sigma_A}(S_2/R_0)}\}_G = \left\{ \begin{array}{c} M\,\vec{\Gamma}_{A/R} \\ \vec{\delta}_{A(S_2/R_0)} = \left[\dfrac{d}{dt}\overrightarrow{\sigma_A}(S_2/R_0)\right]_R \end{array} \right\}$$

$$M.\vec{\Gamma}_{A/R} = M.\frac{d}{dt}(r.\dot{\alpha}\,\overrightarrow{y_1}) = M.r.\frac{d}{dt}(\dot{\alpha}\,\overrightarrow{y_1}) = M.r.\left(\frac{d}{dt}(\dot{\alpha})\overrightarrow{y_1} + \dot{\alpha}\frac{d}{dt}(\overrightarrow{y_1})\right)$$

$$M.\vec{\Gamma}_{G/R} = M.r.(\ddot{\alpha}.\overrightarrow{y_1} - \dot{\alpha}^2.\overrightarrow{x_1})$$

$$\vec{\delta}_{A(S_2/R_0)} = \left[\frac{d}{dt}\overrightarrow{\sigma_A}(S_2/R_0)\right]_R = \frac{d}{dt}\left(\frac{m_2.R^2}{2}(\dot{\alpha}+\dot{\beta})\overrightarrow{z_0}\right) = \frac{m_2.R^2}{2}(\ddot{\alpha}+\ddot{\beta})\overrightarrow{z_0}$$

$$\{D(S_2/R_0)\}_A = \left\{ \begin{array}{c} M\,\vec{\Gamma}_{A/R} \\ \vec{\delta}_{A(S_2/R_0)} \end{array} \right\} = \left\{ \begin{array}{c|c} (m_1+m_2)r.[-\ddot{\alpha}.s\alpha - \dot{\alpha}^2.c\alpha] & 0 \\ (m_1+m_2)r.[\ddot{\alpha}.c\alpha - \dot{\alpha}^2.s\alpha] & 0 \\ 0 & \dfrac{m_2.R^2}{2}(\ddot{\alpha}+\ddot{\beta}) \end{array} \right\}$$

It is assumed that <u>the assembly (S_2)</u> is subject to the following mechanical actions:

- actions of gravity at point A : $\{T_{g\to S_2}\}_A = \left\{ \begin{array}{c|c} 0 & 0 \\ -M.g & 0 \\ 0 & 0 \end{array} \right\}_A$

- action of motor $\mathbf{M_2}$ at point A: $\{T_{M_2\to S_2}\}_A = \left\{ \begin{array}{c|c} 0 & 0 \\ 0 & 0 \\ 0 & C_2 \end{array} \right\}_A$

- action of the arm (S_1) at point A : $\{T_{S_1\to S_2}\}_A = \left\{ \begin{array}{c|c} X_1 & L_1 \\ Y_1 & M_1 \\ Z_1 & 0 \end{array} \right\}_A$

- action of the tube (S_3) at point A : $\{T_{S_1\to S_2}\}_A = \left\{ \begin{array}{c|c} T & -\dfrac{N.L}{2} \\ N & \dfrac{T.L}{2} \\ 0 & T.R \end{array} \right\}_A$

8) Deduce, in the base $\mathbf{R_0}$, the torsor associated with these actions at point A: $\{T_{\overline{S_2}\to S_2}\}_A$

$$\{T_{\overline{S_2}\to S_2}\}_A = \{T_{g\to S_2}\}_A + \{T_{S_1\to S_2}\}_A + \{T_{M_2\to S_2}\}_A + \{T_{S_1\to S_2}\}_A$$

$$\{T_{\overline{S_2}\to S_2}\}_A = \left\{ \begin{array}{c|c} 0 & 0 \\ -M_D.g & 0 \\ 0 & 0 \end{array} \right\}_A + \left\{ \begin{array}{c|c} X_1 & L_1 \\ Y_1 & M_1 \\ Z_1 & 0 \end{array} \right\}_A + \left\{ \begin{array}{c|c} 0 & 0 \\ 0 & 0 \\ 0 & C_2 \end{array} \right\}_A + \left\{ \begin{array}{c|c} T & -\dfrac{N.L}{2} \\ N & \dfrac{T.L}{2} \\ 0 & T.R \end{array} \right\}_A$$

$$\{T_{\overline{S_2}\to S_2}\}_A = \left\{ \begin{array}{c|c} X_1 + T & L_1 - \dfrac{N.L}{2} \\ -M.g + Y_1 + N & M_1 + \dfrac{T.L}{2} \\ Z_1 & C_2 + T.R \end{array} \right\}_A$$

9) Determine the scalar equations resulting from the PFD (Fundamental Principle of Dynamics) applied to the assembly (S_2) in motion relative to $\mathbf{R_0}$. Deduce T and N as a function of $\mathbf{C_2}$.

According to the PFD

$$\{D_{\Sigma/Rg}\}_A = \{T_{\overline{S_2}\to S_2}\}_A \quad \text{i.e.} \quad \begin{Bmatrix} m.\vec{\Gamma}_{A/R} \\ \vec{\delta}_{A(S_2/R_0)} \end{Bmatrix}_A = \begin{Bmatrix} \vec{R}_{(\overline{S_2}\to S_2)} \\ \vec{M}_{(\overline{S_2}\to S_2)} \end{Bmatrix}_A$$

$$\begin{Bmatrix} (m_1+m_2)r.[-\ddot{\alpha}.s\alpha.-\dot{\alpha}^2.c\alpha] \\ (m_1+m_2)r.[\ddot{\alpha}.c\alpha.-\dot{\alpha}^2.s\alpha] \\ 0 \end{Bmatrix}\begin{matrix} 0 \\ 0 \\ \frac{m_2.R^2}{2}(\ddot{\alpha}+\ddot{\beta}) \end{matrix} = \begin{Bmatrix} X_1+T \\ -(m_1+m_2).g+Y_1+N \\ Z \end{Bmatrix}\begin{matrix} L_1-\frac{N.L}{2} \\ M_1+\frac{T.L}{2} \\ C_2+T.R \end{matrix}_A$$

$$\begin{cases} (m_1+m_2)r.[-\ddot{\alpha}.s\alpha.-\dot{\alpha}^2.c\alpha] = X_1+T \\ (m_1+m_2)r.[\ddot{\alpha}.c\alpha.-\dot{\alpha}^2.s\alpha] = -(m_1+m_2).g+Y_1+N \\ 0 = Z_1 \\ 0 = L_1-\frac{N.L}{2} \\ 0 = M_1+\frac{T.L}{2} \\ \frac{m_2.R^2}{2}(\ddot{\alpha}+\ddot{\beta}) = C_2+T.R \end{cases}$$

$$T = \frac{1}{R}\left(\frac{m_2.R^2}{2}(\ddot{\alpha}+\ddot{\beta})-C_2\right)$$

10) Determine the equation of motion of the centre of mass G.

The vector equations of the resultant and the dynamic moment each lead to three scalar equations, i.e. for the two to six scalar equations. In our case we have three equations which are :

$$(m_1+m_2)r.[-\ddot{\alpha}.s\alpha.-\dot{\alpha}^2.c\alpha] = X_1+T \qquad \Rightarrow \qquad \ddot{\alpha}.s\alpha.+\dot{\alpha}^2.c\alpha+\frac{X_1+T}{(m_1+m_2)r} = 0$$

$$(m_1+m_2)r.[\ddot{\alpha}.c\alpha.-\dot{\alpha}^2.s\alpha] = -(m_1+m_2).g+Y_1+N \qquad \Rightarrow \qquad \ddot{\alpha}.c\alpha.-\dot{\alpha}^2.s\alpha+$$

$$\frac{(m_1+m_2).g-Y_1-N}{(m_1+m_2)r} = 0$$

$$\frac{m_2.R^2}{2}(\ddot{\alpha}+\ddot{\beta}) = C_2+T.R \qquad \Rightarrow \qquad \ddot{\alpha}+\ddot{\beta}-\frac{C_2+T.R}{\frac{m_2.R^2}{2}} = 0$$

6. Artisanal oil grinder

6.1. Statement

Figure (1) shows an **artisanal oil millstone** used to crush and grind olives before the dewatering stage.

To simplify <u>the modelling of</u> the mechanism, as illustrated in Figure 2, the following assumptions are made:

- The **grinding wheel (2)**, of thickness **h**, radius **R** and mass **m**, is rotated by the **arm (1)**.
- The weight and inertia of the **arm (1)** are negligible compared with those of the **grinding wheel (2).**
- The motor **M** exerts a torque C_m on the **arm (1)**.

- The mechanical connections are perfect except for the connection between the **grinding wheel (2)** and the **base (0)**, which has a coefficient of friction **f.**

- The **grinding wheel (2)** rolls without slipping at point **I** on the base.

To set the parameters of the mechanism we consider :

the reference frame R_0 $(A, \overrightarrow{x_0}, \overrightarrow{y_0}, \overrightarrow{z_0})$ is associated with the **base (0)** where $(A, \overrightarrow{z_0})$ is the axis of rotation. The reference frame R_1 $(A, \overrightarrow{x_1}, \overrightarrow{y_1}, \overrightarrow{z_0})$ is associated with the **arm (1)**, where $\alpha = (\overrightarrow{x_0}, \overrightarrow{x_1}) = (\overrightarrow{y_0}, \overrightarrow{y_1})$ (**α is negative**). The reference frame R_2 $(G, \overrightarrow{x_1}, \overrightarrow{y_2}, \overrightarrow{z_2})$ is associated with the **grinding wheel (2)** with $\theta = (\overrightarrow{y_1}, \overrightarrow{y_2}) = (\overrightarrow{z_0}, \overrightarrow{z_2})$.

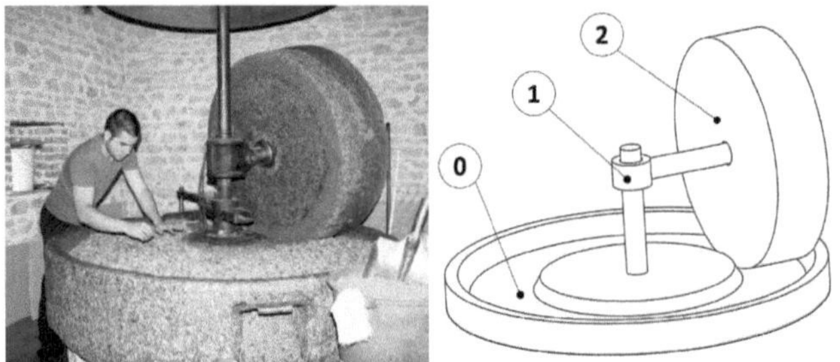

Figure 1: traditional oil millstone

We give :

$\overrightarrow{AG} = a\,\overrightarrow{x_1}$ points G and I lie in the median plane of the wheel.

The inertia matrix of the grinding wheel at point G (the wheel's centre of gravity) is :

$$[I_{(2)}]_G = \begin{bmatrix} A & 0 & 0 \\ 0 & B & 0 \\ 0 & 0 & B \end{bmatrix} \text{ with } A = \frac{m.R^2}{2} \text{ and } B = \frac{m.R^2}{4} + \frac{m.h^2}{12}$$

The static torsors of the mechanical actions are :

$$\{T_{1 \to 2}\} = \begin{Bmatrix} X_1 & 0 \\ Y_1 & M_1 \\ Z_1 & N_1 \end{Bmatrix}_{(G,\overrightarrow{x_1},\overrightarrow{y_1},\overrightarrow{z_0})} ; \{T_{pes \to 2}\} = \begin{Bmatrix} 0 & 0 \\ 0 & 0 \\ -mg & 0 \end{Bmatrix}_{(G,\overrightarrow{x_1},\overrightarrow{y_1},\overrightarrow{z_0})} ; \{T_{0 \to 2}\} = \begin{Bmatrix} 0 & 0 \\ T & 0 \\ N & 0 \end{Bmatrix}_{(I,\overrightarrow{x_1},\overrightarrow{y_1},\overrightarrow{z_0})}$$

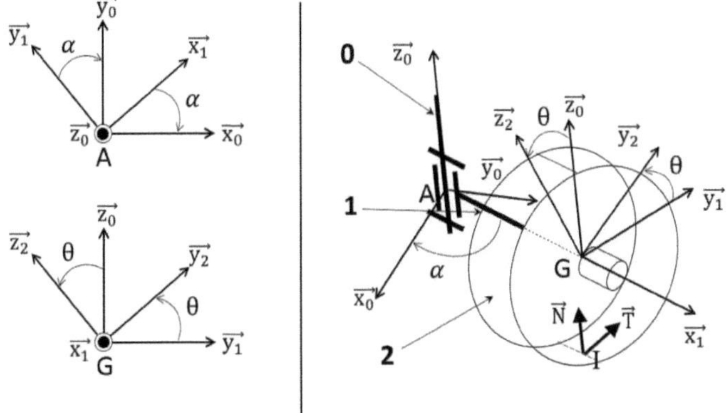

Figure 2: model of the oil grinding wheel

1) Calculate $\vec{\Omega}(R_2/R_1)$ and $\vec{\Omega}(R_1/R_0)$ and deduce $\vec{\Omega}(R_2/R_0)$

2) Show that the speed of the grinding wheel $\vec{V}(G \in 2/R_0) = -a.\dot{\alpha}\,\vec{y_1}$ and Calculate the acceleration $\vec{\Gamma}(G \in 2/R_0)$

3) Show that the angular momentum of the grinding wheel at point G is $\vec{\sigma}_{G(2/R_0)} = A\dot{\theta}\vec{x_1} - B\dot{\alpha}\vec{z_0}$ then write the kinetic torsor of the grinding wheel (2) at point G.

4) Determine the dynamic torsor of the grinding wheel (2) at point G.

5) Apply the PFD to the grinding wheel (2) and write down the equations derived from this principle.

6) deduce the unknowns X_1, Y_1, Z_1, M_1, N_1 and N as a function of the problem data.

6.2. Correction

1) Calculate $\vec{\Omega}(R_2/R_1)$ and $\vec{\Omega}(R_1/R_0)$ and deduce $\vec{\Omega}(R_2/R_0)$

$\vec{\Omega}(R_2/R_1) = \dot{\theta}\vec{x_1}$; $\vec{\Omega}(R_1/R_0) = -\dot{\alpha}\vec{z_0}$; $\vec{\Omega}(R_2/R_0) = \vec{\Omega}(R_2/R_1) + \vec{\Omega}(R_1/R_0) = \dot{\theta}\vec{x_1} - \dot{\alpha}\vec{z_0}$

2) Show that the speed of the grinding wheel $\vec{V}(G \in 2/R_0) = -a.\dot{\alpha}\,\vec{y_1}$ and Calculate the acceleration $\vec{\Gamma}(G \in 2/R_0)$

$$\vec{V}_{(G\in2/0)} = \frac{d\vec{AG}}{dt}\bigg)_{R_0} \overset{Bour}{=} \frac{d\vec{AG}}{dt}\bigg)_{R_1} + \vec{\Omega}(R_1/R_0)\wedge\vec{AG} = -\dot{\alpha}\vec{z_0} \wedge a\,\vec{x_1}$$

Because $\vec{AG} = a\,\vec{x_1}$ is expressed in the B basis$_1$ of R$_1$

$$\vec{V}(G \in 2/R_0) = -a.\dot{\alpha}\,\vec{y_1}$$

$$\vec{\Gamma}_{(G\in2/R_0)} = \frac{d\vec{V}_{(G\in2/R_0)}}{dt}\bigg)_{R_0} \overset{Bour}{=} \frac{d\vec{V}_{(G\in2/R_0)}}{dt}\bigg)_{R_1} + \vec{\Omega}(R_1/R_0) \wedge \vec{V}_{(G\in2/R_0)}$$

$$\vec{\Gamma}_{(G\in2/R_0)} = -a.\ddot{\alpha}\vec{y_1} - a.\dot{\alpha}^2\vec{x_1}$$

3) Show that the angular momentum of the grinding wheel at point G is $\vec{\sigma}_{G(2/R_0)} = A\dot\theta\vec{x_1} - B\dot\alpha\vec{z_0}$ then write the kinetic torsor of the grinding wheel (2) at point G.

$$\vec{\sigma_G}(2/R_0) = [I_G(2)].\vec{\Omega}_{2/R_0} = \begin{bmatrix} A & 0 & 0 \\ 0 & B & 0 \\ 0 & 0 & B \end{bmatrix}\begin{pmatrix} \dot\theta \\ 0 \\ -\dot\alpha \end{pmatrix} = A\dot\theta\vec{x_1} - B\dot\alpha\vec{z_0}$$

$$\{C_{(2/R_0)}\}_G = \left\{ \begin{array}{c} m\,\vec{V}_{(G\in 2/R_0)} \\ \vec{\sigma}_{G(2/R_0)} \end{array} \right\} = \left\{ \begin{array}{c} \vec{Q}_{2/R_0} = -m.a.\dot\alpha\,\vec{y_1} \\ \vec{\sigma}_{G(2/R_0)} = A\dot\theta\vec{x_1} - B\dot\alpha\vec{z_0} \end{array} \right\}$$

4) Determine the dynamic torsor of the grinding wheel (2) at point G.

$$\{D_{(2/R_0)}\}_G = \left\{ \begin{array}{c} m\,\vec{\Gamma}_{(G\in 2/R_0)} \\ \vec{\delta}_{G(2/R_0)} \end{array} \right\}$$

$$\vec{\delta_G}(2/R_0) = \left[\frac{d}{dt}\vec{\sigma_A}(2/R_0)\right]_R = A\ddot\theta\vec{x_1} - A\dot\theta.\dot\alpha\vec{y_1} - B\ddot\alpha\vec{z_0}$$

$$\{D_{(2/R_0)}\}_G = \left\{ \begin{array}{c} m\,\vec{\Gamma}_{(G\in 2/R_0)} = m.(-a.\ddot\alpha\vec{y_1} - a.\dot\alpha^2\vec{x_1}) \\ \vec{\delta}_{G(2/R_0)} = A\ddot\theta\vec{x_1} - A\dot\theta.\dot\alpha\vec{y_1} - B\ddot\alpha\vec{z_0} \end{array} \right\}$$

5) Apply the Fundamental Principle of Dynamics (FPD) to the grinding wheel (2) and write down the equations derived from this principle.

The static torsors of the mechanical actions are :

$$\{T_{1\to 2}\} = \left\{ \begin{array}{cc} X_1 & 0 \\ Y_1 & M_1 \\ Z_1 & N_1 \end{array} \right\}_{(G,\vec{x_1},\vec{y_1},\vec{z_0})} ; \{T_{pes\to 2}\} = \left\{ \begin{array}{cc} 0 & 0 \\ 0 & 0 \\ -mg & 0 \end{array} \right\}_{(G,\vec{x_1},\vec{y_1},\vec{z_0})} ; \{T_{0\to 2}\} = \left\{ \begin{array}{cc} 0 & 0 \\ T & 0 \\ N & 0 \end{array} \right\}_{(I,\vec{x_1},\vec{y_1},\vec{z_0})}$$

$$\{D_{(2/R_0)}\}_G = \{T_{\bar 2 \to 2}\}_G = \{T_{1\to 2}\}_G + \{T_{pes\to 2}\}_G + \{T_{0\to 2}\}_G$$

$$\left\{ \begin{array}{cc} -m.a.\dot\alpha^2 & A\ddot\theta \\ -m.a.\ddot\alpha & -A\dot\theta.\dot\alpha \\ 0 & -B\ddot\alpha \end{array} \right\} = \left\{ \begin{array}{cc} X_1 & 0 \\ Y_1 & M_1 \\ Z_1 & N_1 \end{array} \right\} + \left\{ \begin{array}{cc} 0 & 0 \\ 0 & 0 \\ -mg & 0 \end{array} \right\} + \left\{ \begin{array}{cc} 0 & R.T \\ T & 0 \\ N & 0 \end{array} \right\}$$

6) deduce the unknowns X_1 , Y_1 , Z_1 , M_1 , N_1 and N as a function of the data in the problem0

$$\left\{ \begin{array}{l} X_1 = -m.a.\dot\alpha^2 \\ Y_1 = -m.a.\ddot\alpha - T \\ Z_1 = mg - N \end{array} \right. \qquad \left\{ \begin{array}{l} R.T = A\ddot\theta \\ M_1 = -A\dot\theta.\dot\alpha \\ N_1 = -B\ddot\alpha \end{array} \right.$$

7. Grinding machine

7.1. Statement

The grinder (Σ) shown in Figures 1 and 2 consists of the following four solids:

A frame (S0), two identical grinding wheels (S1) and (S3): of thickness e_1 , radius R_1 , densityρ_1 , centres of mass G_1 and G_3 linked to the reference frame R_1 $(G_1, \vec{x_1}, \vec{y_1}, \vec{z_0})$ rotating relative to a fixed reference frame R_0 $(O, \vec{x_0}, \vec{y_0}, \vec{z_0})$ along the axis $\vec{z_0}$.

A uniform shaft (S2) of length L_2 , radius R_2 , densityρ_2 and centre of mass **O** linked to the reference frame R_1 $(G_1, \vec{x_1}, \vec{y_1}, \vec{z_0})$. It is pivotally connected by axis $(G_2, \vec{z_0})$ in relation to the frame (S0) with $\theta = (\vec{x_0}, \vec{x_1}) = (\vec{y_0}, \vec{y_1})$.

It rotates at nominal speed N = 2850 rpm. To reach this speed, it takes DT1=1.5 s (the initial rotation speed being zero). When the motor power supply is cut off, the spindle takes DT2= 40 s to stop.

Figure 1: Grinder system

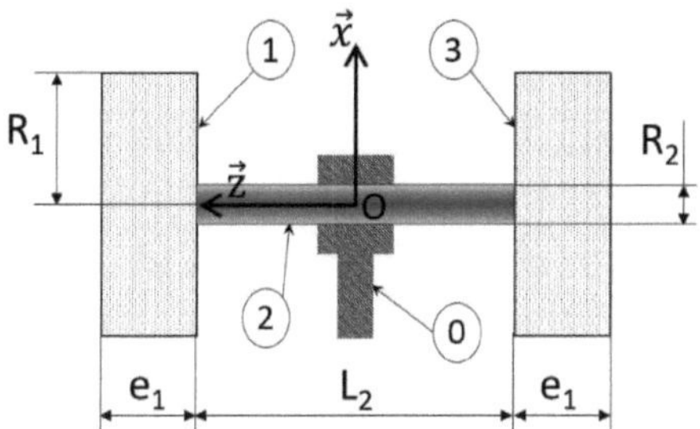

Figure 2: Modelling a system

1) Give the co-ordinates of the centres of mass (of inertia) $(G_1$, G_2 and G_3) of each part of the grinding machine $(\Sigma$) in the reference frame $(O, \vec{x_0}, \vec{y_0}, \vec{z_0})$.

2) Show that **(O)** is the position of the centre of mass of the grinding wheel assembly (Σ).

3) Determine the inertia matrix of a cylinder of length **L**, radius **R** and axis $\vec{z_0}$expressed at the centre of inertia **G** is of the form :

4) Give the expressions for the inertia matrices $[I_{G1}\,(S1)]$, $[I_O\,(S2)]$ and $[I_{G3}\,(S3)]$

5) Find the inertia matrix of the grinding wheel $[I_O\,(\Sigma)]$ expressed at point (O)

6) Determine the kinetic torsor of the grinding wheel $(\Sigma$) at point O

7) Determine the dynamic torsor $\{D_{(\Sigma/R_0)}\}_O$ of the grinding wheel $(\Sigma$) at point O

Let $\{T_{\bar{\Sigma} \to \Sigma}\}_O$ is the static torsor of the external actions applied to the grinding machine $(\Sigma$) expressed at point O.

$$\{T_{\Sigma \to \Sigma}\}_O = \begin{Bmatrix} X_1 & L_1 \\ Y_1 & M_1 \\ Z_1 & 0 \end{Bmatrix} + \begin{Bmatrix} 0 & 0 \\ 0 & 0 \\ 0 & C \end{Bmatrix} + \begin{Bmatrix} -P1 - P2 - P3 & 0 \\ 0 & 0 \\ 0 & 0 \end{Bmatrix}$$

Where **C** is the torque applied by a motor to the shaft (S2).

8) Apply the fundamental principle of dynamics ($\{T_{\Sigma \to \Sigma}\}_O = \{D_{(\Sigma/R_0)}\}_O$) to the grinding wheel (Σ) at point O and write down the equations derived from this principle. Then show that the

torque $C = \left(m_1.R_1{}^2 + \frac{m_2.R_2{}^2}{2} \right) \ddot{\theta}$ ($\ddot{\theta}$ is the angular acceleration)

7.2.Correction

1) Give the co-ordinates of the centres of mass (of inertia) (G_1 , G_2 and G_3) of each part of the grinding machine (Σ) in the reference frame ($O, \overrightarrow{x_0}, \overrightarrow{y_0}, \overrightarrow{z_0}$).

$G_1\left(0, 0, \frac{L+e}{2}\right)$; $G_2(0, 0, 0)$; $G_3\left(0, 0, -\frac{L+e}{2}\right)$

2) Show that **(O)** is the position of the centre of mass of the grinding wheel assembly ( ).Σ

Let G be the centre of mass of the system

$M.\overrightarrow{OG} = \sum_{i=1}^{n} \overrightarrow{OG_i}.m_i$ and $M = \sum_{i=1}^{n} m_i$

$M.\overrightarrow{OG} = m_1.\overrightarrow{OG_1} + m_2.\overrightarrow{OG_2} + m_1.\overrightarrow{OG_3}$ with $M = m_1 + m_2 + m_1$

We have $\overrightarrow{OG_2} = \vec{0}$ so

$$M.\overrightarrow{OG} = m_1.\overrightarrow{OG_1} + m_1.\overrightarrow{OG_3} = m_1 \begin{pmatrix} 0 \\ 0 \\ \frac{L+e}{2} \end{pmatrix} + m_1 \begin{pmatrix} 0 \\ 0 \\ -\frac{L+e}{2} \end{pmatrix} = \vec{0}$$

Point G therefore coincides with point O

3) Determine the inertia matrix of a cylinder of length **L**, radius **R** and axis $\overrightarrow{z_0}$ expressed at the centre of inertia **G**.

The axis $\overrightarrow{z_0}$ axis is an axis of revolution, so all the products of inertia are zero.

The axes $(o, \vec{x})$ and $(o, \vec{y})$ are equivalent, so $A = B$

$$[I_{(G,\vec{x},\vec{y},\vec{z})}(S)] = \begin{bmatrix} A & 0 & 0 \\ 0 & A & 0 \\ 0 & 0 & C \end{bmatrix}$$

$A = \int (x^2 + z^2).dm \quad (1)$

$A = \int (y^2 + z^2).dm \quad (2)$

$C = \int (x^2 + y^2).dm \quad (3)$

Let's add member to member the relations (1) and (2)

$2.A = \int (x^2 + y^2).dm + 2.\int z^2.dm$

This reveals C. from which

$A = \frac{C}{2} + \int z^2 . dm$ with $dm = \rho . r . dr . d\theta . dz$

$dm = \rho . \pi . R^2 . dz = \frac{m}{h} dz$ hence $\int z^2 . dm = \frac{m}{h} . \int z^2 . dz = \frac{m.L^2}{12}$

$C = \int (x^2 + y^2) . dm = \int r^2 . dm = \frac{2.m}{R^2} \int r^3 . dr = \frac{m.R^2}{2}$

$A = \frac{C}{2} + \int z^2 . dm = \frac{m.R^2}{4} + \frac{m.L^2}{12}$

4) Give the expressions for the inertia matrices $[I_{G_1} (S1)]$, $[I_O (S2)]$ and $[I_{G3} (S3)]$

$$[I_{(G_1)}(S1)] = \begin{bmatrix} \frac{m_1.R_1{}^2}{4} + \frac{m_1.e_1{}^2}{12} & 0 & 0 \\ 0 & \frac{m_1.R_1{}^2}{4} + \frac{m_1.e_1{}^2}{12} & 0 \\ 0 & 0 & \frac{m_1.R_1{}^2}{2} \end{bmatrix}$$

$[I_{(O)}(S2)] =$

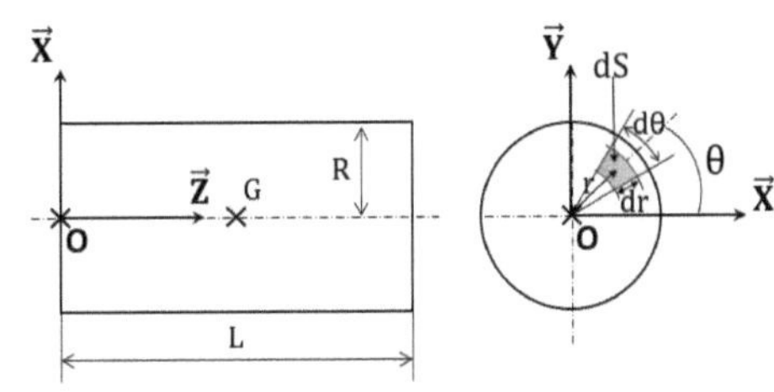

$$\begin{bmatrix} \frac{m_2.R_2{}^2}{4} + \frac{m_2.L_2{}^2}{12} & 0 & 0 \\ 0 & \frac{m_2.R_2{}^2}{4} + \frac{m_2.L_2{}^2}{12} & 0 \\ 0 & 0 & \frac{m_2.R_2{}^2}{2} \end{bmatrix}$$

$$[I_{(G_3)}(S3)] = \begin{bmatrix} \frac{m_1.R_1{}^2}{4} + \frac{m_1.e_1{}^2}{12} & 0 & 0 \\ 0 & \frac{m_1.R_1{}^2}{4} + \frac{m_1.e_1{}^2}{12} & 0 \\ 0 & 0 & \frac{m_1.R_1{}^2}{2} \end{bmatrix}$$

5) Find the inertia matrix of the grinding wheel $[I_O (\Sigma)]$ expressed at point (O)

$[I_O (\Sigma)] = [I_O (S1)] + [I_O (S2)] + [I_O (S3)]$

To determine the expression of the matrix $[I_{G_1} (S1)]$ at point O we use the Huygens theorem

$[I_{(O)}(S1)] = [I_{(G_1)}(S1)] + [I(G.O.m)]$

$$[I_{(O)}(S1)] = \begin{bmatrix} \frac{m_1.R_1^2}{4} + \frac{m_1.e_1^2}{12} & 0 & 0 \\ 0 & \frac{m_1.R_1^2}{4} + \frac{m_1.e_1^2}{12} & 0 \\ 0 & 0 & \frac{m_1.R_1^2}{2} \end{bmatrix} + m_1 \begin{bmatrix} \left(\frac{L+e}{2}\right)^2 & 0 & 0 \\ 0 & \left(\frac{L+e}{2}\right)^2 & 0 \\ 0 & 0 & 0 \end{bmatrix}$$

To determine the expression of the matrix $\left[I_{G_3}(S3)\right]$ at point O we use the Huygens theorem

$$[I_{(O)}(S3)] = \begin{bmatrix} \frac{m_1.R_1^2}{4} + \frac{m_1.e_1^2}{12} & 0 & 0 \\ 0 & \frac{m_1.R_1^2}{4} + \frac{m_1.e_1^2}{12} & 0 \\ 0 & 0 & \frac{m_1.R_1^2}{2} \end{bmatrix} + m_1 \begin{bmatrix} \left(\frac{L+e}{2}\right)^2 & 0 & 0 \\ 0 & \left(\frac{L+e}{2}\right)^2 & 0 \\ 0 & 0 & 0 \end{bmatrix}$$

$$[I_O(\Sigma)] = [I_O(S1)] + [I_O(S2)] + [I_O(S3)]$$

$$[I_O(\Sigma)] = 2.m_1 \begin{bmatrix} \frac{R_1^2}{4} + \frac{e_1^2}{12} + \left(\frac{L+e}{2}\right)^2 & 0 & 0 \\ 0 & \frac{R_1^2}{4} + \frac{e_1^2}{12}\left(\frac{L+e}{2}\right)^2 & 0 \\ 0 & 0 & \frac{R_1^2}{2} \end{bmatrix} +$$

$$m_2 \begin{bmatrix} \frac{R_2^2}{4} + \frac{L_2^2}{12} & 0 & 0 \\ 0 & \frac{R_2^2}{4} + \frac{L_2^2}{12} & 0 \\ 0 & 0 & \frac{R_2^2}{2} \end{bmatrix}$$

6) Give the kinetic torsor $\{C_{(\Sigma/R_0)}\}_O$.

$$\vec{\sigma_O}(\Sigma/R_0) = [I_O(\Sigma)].\vec{\Omega}(\Sigma/R_0) = \begin{bmatrix} A & 0 & 0 \\ 0 & A & 0 \\ 0 & 0 & C \end{bmatrix}.\begin{pmatrix} 0 \\ 0 \\ \dot{\theta} \end{pmatrix}$$

$$A = 2.m_1\left[\frac{R_1^2}{4} + \frac{e_1^2}{12} + \left(\frac{L+e}{2}\right)^2\right] + m_2\left[\frac{R_2^2}{4} + \frac{L_2^2}{12}\right] \text{ and } C = 2.\frac{m_1.R_1^2}{2} + \frac{m_2.R_2^2}{2}$$

$$\vec{\sigma_O}(\Sigma/R_0) = \left(2.\frac{m_1.R_1^2}{2} + \frac{m_2.R_2^2}{2}\right)\dot{\theta}\,\vec{Z}$$

the kinetic torsor is

$$\{C_{(\Sigma/R_0)}\}_O = \left\{ \begin{array}{l} \vec{Q} = M\,\vec{V}_{B\in S2/R0} \\ \vec{\sigma_O}(\Sigma/R_0) = [I_O(\Sigma)].\vec{\Omega}(\Sigma/R_0)) \end{array} \right\} = \left\{ \begin{array}{l|l} 0 & 0 \\ 0 & 0 \\ 0 & \left(2.\frac{m_1.R_1^2}{2} + \frac{m_2.R_2^2}{2}\right)\dot{\theta} \end{array} \right\}$$

With $\vec{\Omega}(\Sigma/R_0) = \dot{\theta}\,\vec{Z}$ is the speed of rotation of the system relative to the reference frame R_0

7) Determine the dynamic torsor $\{D_{(\Sigma/R_0)}\}_O$ of the grinding wheel (Σ) at point O

$$\{D_{(\Sigma/R_0)}\}_O = \left\{ \begin{array}{c} M\,\vec{\Gamma}_{A/R} \\ \vec{\delta}_{A(S_2/R_0)} = \left[\dfrac{d}{dt}\overrightarrow{\sigma_A}(S_2/R_0)\right]_R \end{array} \right\}$$

$$\vec{\delta}_{O(\Sigma/R_0)} = \left[\dfrac{d}{dt}\overrightarrow{\sigma_O}(\Sigma/R_0)\right]_R = \dfrac{d}{dt}\left(\left(2.\dfrac{m_1.R_1{}^2}{2} + \dfrac{m_2.R_2{}^2}{2}\right)\dot{\theta}\,\overrightarrow{z_0}\right) = \left(2.\dfrac{m_1.R_1{}^2}{2} + \dfrac{m_2.R_2{}^2}{2}\right)\ddot{\theta}\,\overrightarrow{z_0}$$

$$\{D_{(\Sigma/R_0)}\}_O = \left\{ \begin{array}{c} M\,\vec{\Gamma}_{O/R} \\ \vec{\delta}_{O(\Sigma/R_0)} \end{array} \right\} = \left\{ \begin{array}{c|c} 0 & 0 \\ 0 & 0 \\ 0 & \left(2.\dfrac{m_1.R_1{}^2}{2} + \dfrac{m_2.R_2{}^2}{2}\right)\ddot{\theta} \end{array} \right\}$$

8) Apply the PFD to the grinding wheel (Σ) at point O and write down the equations derived from this principle.

Let $\{T_{\overline{\Sigma}\to\Sigma}\}_O$ is the static torsor of the external actions applied to the grinding machine (Σ) expressed at point O.

$$\{T_{\overline{\Sigma}\to\Sigma}\}_O = \left\{ \begin{array}{c|c} X_1 & L_1 \\ Y_1 & M_1 \\ Z_1 & 0 \end{array} \right\} + \left\{ \begin{array}{c|c} 0 & 0 \\ 0 & 0 \\ 0 & C \end{array} \right\} + \left\{ \begin{array}{c|c} -P1 - P2 - P3 & 0 \\ 0 & 0 \\ 0 & 0 \end{array} \right\} = \left\{ \begin{array}{c|c} X_1 - P1 - P2 - P3 & L_1 \\ Y_1 & M_1 \\ Z_1 & C \end{array} \right\}$$

Where **C** is the torque applied by a motor to the shaft (S2).

According to the PFD

$$\{D_{(\Sigma/R_0)}\}_A = \{T_{\overline{S_2}\to S_2}\}_A \quad \text{i.e.} \quad \left\{ \begin{array}{c} m.\vec{\Gamma}_{O/R} \\ \vec{\delta}_{O(\Sigma/R_0)} \end{array} \right\}_O = \left\{ \begin{array}{c} \vec{R}_{(\overline{\Sigma}\to\Sigma)} \\ \vec{M}_{(\overline{\Sigma}\to\Sigma)} \end{array} \right\}_O$$

$$\left\{ \begin{array}{c|c} 0 & 0 \\ 0 & 0 \\ 0 & \left(2.\dfrac{m_1.R_1{}^2}{2} + \dfrac{m_2.R_2{}^2}{2}\right)\ddot{\theta} \end{array} \right\} = \left\{ \begin{array}{c|c} X_1 - P1 - P2 - P3 & L_1 \\ Y_1 & M_1 \\ Z_1 & C \end{array} \right\}$$

$Y_1 = Z_1 = 0$

$L_1 = M_1 = 0$

Engine torque is $C = \left(2.\dfrac{m_1.R_1{}^2}{2} + \dfrac{m_2.R_2{}^2}{2}\right)\ddot{\theta}$

References

[1] Yves Gourinat, *exercices et problèmes de mécanique des solides et des structures*, dunod, paris, 2011 ISBN 978-2-10-057122 -2

[2] Pierre Agati (author), Yves Brémont, Gérard Delville, *Mécanique du solide - applications industrielles*, dunod, paris, 1996

[3] Mohamed SOULA Habib SAHLAOUI Amara NEJI mécanique des systèmes de solides cours et exercices corrigés, centre de publication universitaire,Tunis, 2012

[4] Mahfoudh AYADI and Atef BOULILA, Dynamics of solid systems , Centre de publication universitaire, 2017

[5] Sylvie Pommier, Yves Berthaud, *Mécanique générale cours et exercices corrigés*, dunod, paris, 2010 ISBN 978-2-10-054820-0

[6] Queyrel, J.-L., Mesplede, J., Précis de Physique Mécanique MPSI Cours - Méthodes - Exercices résolus, Bréal, Paris, 1995, ISBN: 978285394787

[7] Note de cours - Ali kadi *mécanique rationnelle cours & exercices résolus* université Mohamed Bougara - Boumerdes
 https://cours-examens.org/images/Etudes_superieures/TCT_2_annee/Mecanique_rationnelle/1mecanique_rationnelle_book.pdf

[8] Note de cours 2014 - M. Bourich *Copies of exercises and exams: Mechanics of Undeformable Solid Systems* ENSA de Marrakech
http://www.ensa.ac.ma/docs/pedagogie/MecDesSysSolIndef_Polycop_Ex.pdf

[9] Note de cours - 2010 Pascal Chambraud *Cours - 2ème année Mécanique, Automatique* Lycée Hoche 73 avenue de Saint-Cloud 78000 Versailles
http://back.maquisdoc.net/data/tex_pascal/prepas_2.tex/prepas_2.pdf

[10] Doina Dragulescu, mécanique générale et applications, L'harmattan, 2005, ISBN: 2-7475-9036-4

[11] Claude Chèze, Hélène Lange, Mécanique générale - 1er et 2e cycles Cours, exercices et problèmes corrigés, Ellipses, 1998, ISBN: 9782729845179

Printed by Books on Demand GmbH, Norderstedt / Germany